Introduction to
Sediment
Geochemistry

——Introduction to——
Sediment
Geochemistry

Sergei Katsev

University of Minnesota Duluth, USA

World Scientific

NEW JERSEY · LONDON · SINGAPORE · BEIJING · SHANGHAI · HONG KONG · TAIPEI · CHENNAI · TOKYO

Published by

World Scientific Publishing Co. Pte. Ltd.

5 Toh Tuck Link, Singapore 596224

USA office: 27 Warren Street, Suite 401-402, Hackensack, NJ 07601

UK office: 57 Shelton Street, Covent Garden, London WC2H 9HE

Library of Congress Control Number: 2025042210

British Library Cataloguing-in-Publication Data
A catalogue record for this book is available from the British Library.

Cover photo by Sharon Dawson.

INTRODUCTION TO SEDIMENT GEOCHEMISTRY

ISBN 978-981-98-2126-6 (hardcover)
ISBN 978-981-98-2127-3 (ebook for institutions)
ISBN 978-981-98-2128-0 (ebook for individuals)

For any available supplementary material, please visit
https://www.worldscientific.com/worldscibooks/10.1142/14527#t=suppl

Desk Editors: Eshak Nabi Akbar Ali/Amanda Yun

Typeset by Stallion Press
Email: enquiries@stallionpress.com

To the memory of my mother

Preface

The idea of writing this textbook first arose when I started training graduate students and realized I needed a short book to introduce the subject of sediment early diagenesis. Canonical books on the subject tended to be formidably thick, filling hundreds of pages with many details. It also did not help that most of these books focused on marine sediments, whereas my own research dealt mainly with freshwater lakes, which seemed to have been largely left out. This book is an attempt to remedy this situation. It is not meant to be exhaustive but tries to cover the essentials of sediment geochemistry succinctly, to serve as a quick yet comprehensive introduction. The book covers both theoretical and practical aspects, and contains practically relevant exercises that students can use to test their understanding.

Many ideas for this book were developed while teaching a course on sediment geochemistry at the University of Minnesota Duluth, and especially as a result of team-teaching a graduate course on introductory limnology. Limnology – the science of lakes – is an interdisciplinary field, so students in these courses came from a variety of backgrounds spanning Geology, Chemistry, Physics, and Biology. Teaching such a diverse audience required the instructors to provide a lot of background information that was not always included in more specialized textbooks. This book tries to incorporate such background information throughout the text and also in the Appendices.

I am grateful and indebted to a large number of people with whom I had the privilege to discuss the ideas in this book and to learn from: my colleagues at the Large Lakes Observatory, especially Bob Sterner, Katie Schreiner, Byron Steinman, and Ted Ozersky; students in my courses and graduate students and postdocs in my lab, especially Jiying Li. Over the years, my understanding of sediment geochemistry was shaped by interactions with Alfonso Mucci, Maria Dittrich, Sean Crowe, Christof Meile, Itay Halevy, Ivan L'Heureux, Denis Rancourt, and many others. Maria Dittrich has done an admirable and selfless job of reading through the draft of this book and making numerous helpful suggestions. My thinking, and I hope my writing style, have been profoundly influenced by my interactions with my postdoctoral mentor Bjorn Sundby, whose wisdom shaped the way I now see this field. My writing of this book has benefited from the previous works on the same topics, particularly from the books by Robert Berner [1], Bernard Boudreau [2], David Burdige [3], Kurt Konhauser [4], Craig Bethke [5], and Donald Canfield, Bo Thamdrup and Erik Kristensen [6]. These books cover their subjects in much more detail than it was possible in this short course, and an interested reader is well advised to consult them for further information.

Finally, the list of acknowledgements would be painfully incomplete without mentioning the best-intentioned interventions of my family, without whom this book would have been written much faster but the process wouldn't be nearly as fun.

About the Author

Sergei Katsev is a Professor at the University of Minnesota Duluth's Large Lakes Observatory, an interdisciplinary institution dedicated to the studies of large lakes worldwide. He holds a joint appointment at the Department of Physics and Astronomy and is a member of the University of Minnesota's graduate program in Water Resources Science. He teaches physics and limnology and conducts research at the intersection of hydrodynamics, geochemistry, and geomicrobiology.

Contents

Chapter 1

Introduction

1.1 Why Sediments

At first glance, it may appear to be an odd choice to write a book on the science of the muddy deposits that cover the bottoms of oceans and lakes. After all, most of the aquatic processes that are of common interest happen in the water column: the energy of sunlight is harnessed by photosynthesis, phytoplankton grow and die, zooplankton feed on phytoplankton, fish consume both, and humans catch fish. Geologists, of course, pay much attention to the sedimentary rocks into which sediments transform after millions of years. A typical course in sedimentary geology, however, does not spend much time on the uppermost, unconsolidated sediments. Yet, many geochemists need to focus on locations where reactions take place at high rates, catalyzed by high numbers of microbial cells – in the upper few centimeters of sediments. The seemingly unremarkable pile of debris at the bottom of a water body is effectively a bioreactor, an incinerator that burns a large fraction of the deposited organic material, a recycling factory that returns essential nutrients, and a storage facility for toxic wastes and the precious records of environmental history. Confined to the upper centimeters of fluid-filled sediments, this reactive layer serves as a boundary between the water that makes up the hydrosphere and the sedimentary rocks that belong to the lithosphere. While all the processes that turn the deposited material into (eventually) a sedimentary rock are known collectively as *diagenesis*, this book concerns itself with the transformations that happen while the sediment is still actively exchanging substances with the overlying water, the *early diagenesis*.

The science of (early diagenetic) sediment geochemistry finds itself at the intersection of several disciplines that rarely overlap in other contexts. At one end there is Ecology. It looks into the processes in aquatic systems that are mostly biological in nature and happen over the course of months, years or decades. Chemical conditions naturally have strong effects on them. Phosphorus, for example, is a nutrient that is needed by all organisms, so its scarcity commonly limits algal growth in inland lakes. Attempts to limit the amounts of phosphorus in lakes to prevent eutrophication, however, led to the realization that significant quantities of phosphorus were coming from sediments, rather than from watersheds. The sedimentary cycling (and recycling) of phosphorus then became a hot topic of studies and the target of environmental management interventions. Expensive policies were put in place to curtail fluxes of phosphorus from agricultural lands where it was used in fertilizers. In marine environments, vast swaths of the oceans are limited in their biological productivity by the amounts of biologically available nitrogen, which commonly comes in the forms of nitrate or ammonium. Removal of nitrate into the biologically inert N_2 gas happens in the absence of oxygen, which, in the modern ocean, with the exception of a handful of oxygen minimum zones, happens only in sediments. The sedimentary cycling of nitrogen thus becomes important for the entire geochemical budget of nitrogen in the ocean. Processes such as these make sediment geochemistry a key component of the disciplines that aim to understand aquatic environments: limnology and oceanography.

At the other end of the spectrum of disciplines, there is Geology. It takes the long view of millions and billions of years and uses rocks as its primary evidence. On such long time scales, geochemical processes in sediments acquire truly global importance. The fact that the Earth's atmosphere retains such a highly reactive compound as molecular oxygen is a direct consequence of the burial of organic carbon into marine sediments [7]. During the geological periods when a higher proportion of the deposited carbon was being buried, e.g., as a consequence of ocean anoxia, less oxygen could react with it to form carbon dioxide, and consequently more oxygen could stay in the atmosphere. Similarly, at the dawn of life in the Archean eon over three billion years ago, the coupled sedimentary cycling of sulfur and carbon likely determined the proportion of methane gas that could escape into the atmosphere, where its greenhouse effect could off-set the lower luminosity of the young Sun [8]. Sediments from those times not only regulated the contemporary chemical processes but now serve as nearly the only source of evidence for those processes. Sedimentary archives preserve the environmental records for pale-oceanography and paleolimnology, on time scales that range from several years in modern aquatic systems to millions and billions of years in sedimentary rocks. Such records tell us about the past climates and the histories of aquatic bodies and their ecologies. The geochemical, isotopic, and biomarker signals preserved in sediments are the clues that we interpret to decipher the biogeochemical processes of yore.

Yet, there is also Evolutionary Biology. Most of the geochemical reactions should be called biogeochemical, as they are catalyzed by microbes and their enzymes that developed over the nearly four billion years of Earth history. These evolutionary steps were tightly coupled to the evolution of the Earth's geochemical environments in which the ancestral microorganisms thrived. Complex organisms may have originated in sediments, and their evolution has shaped the trajectory of the Earth system. Charles Darwin's last book [9] was dedicated to the effects of earthworms on the geology of the planet. Much in the same way, organisms that crawl on the seafloor have shaped the chemistry of the Ocean [10]. They have done so by modifying the networks of chemical reactions in the upper centimeters of sediments.

Sediments begin in the water column. Mineral and organic particles flow into the surfaces of lakes and oceans from rivers and are deposited as dust from the atmosphere. Organic remains of aquatic organisms sink to the bottom from the surface. On their downward journey, these particles undergo transformations through chemical reactions, precipitation or dissolution of minerals, and biological recycling through numerous food webs. Eventually, however, they reach the bottom where they remain for a considerable period of time.

What happens to them after deposition is the subject of the highly interdisciplinary field of sediment geochemistry. Physical and biological transport, chemical reactions, and microbial metabolisms transform the deposited material, reshaping it before burial into deeper sediment layers. These processes pull some substances into the sediments from the overlying water, recycle others back, and create or destroy habitats for benthic animals. To understand what controls the physical, chemical and biological reactions and how they interact among themselves and with the environment, one needs to understand the fundamentals of sediment geochemistry. This is the subject of this book.

1.2   How to Read This Book

I imagine most readers of this book will use it either as a textbook for a course or as a reference for background information, which could be consulted whenever a particular topic needs to be refreshed in one's memory. Accordingly, chapters of this book may be read sequentially, in the order in which they are presented here, or randomly, as needed. In the latter case, however, the reader may be prompted to flip pages back and forth to look up certain topics in other chapters, as nearly all chapters are interlinked and certain material needed for understanding of one chapter may be contained within another.

Chapters 2–4 are general in their approach, as they outline the physical, chemical, and biological processes that take place in nearly all sediments. Chapters 5–9 focus on the cycling of individual elements: carbon, nitrogen, metals (iron and manganese), phosphorus, and sulfur. While highly dependent on each other, these cycles are singled out for separate treatment as geochemists often find themselves working on one element at a time. Chapter 10 focuses on the effects of sediment geochemistry on paleolimnological and paleoceanographic records. In principle, it may also be possible to skip the general Chapters 3 or 4, if the reader is familiar with the basics of chemistry or geomicrobiology, and refer to the material in those chapters episodically, when needed.

Chapter 11 is dedicated to mathematical and modeling approaches. While sediment geochemistry used to be perceived as a field that was almost purely experimental, today it would put one at a serious disadvantage to ignore the advances in theory that have been made since the pioneering work of Robert Berner [1]. The use of mathematics, nevertheless, remains strangely limited in mainstream sediment geochemistry, and this book tries to fill that gap. Normally, there would be a chapter dedicated to numerical modeling. In this book, however, I chose to emphasize the mathematical descriptions that do not require a computer or any knowledge of programming. In fact, numerical estimates that require the use of mathematics are scattered throughout the text of most chapters. Chapter 11 therefore can be used not only on its own, but also as an accessory text to other chapters. The mathematical approaches that it presents are general and can be applied to multiple processes and elemental cycles. Perhaps a more didactic approach would be to place this chapter immediately after the "general" Chapters 2–4, or even intersperse its material within them, sprinkling bits of math here and there. I made a choice to place it near the end, however, to make it easier for some of the readers who are not steeped in math. A mathematically inclined reader, however, would be well advised to read this chapter before proceeding to the "elemental cycles" chapters, or consult its content periodically whenever it is referenced in the text.

Each chapter concludes with a set of exercises. Expositions of the solutions to some of them are collected in Chapter 12. These exercises are meant to help the reader check their grasp of the material, and to some degree as an aid to a hypothetical course instructor. In line with the reasoning outlined above, many of the exercises focus on making numerical estimates and applying mathematics. To some degree, this reflects the personal bias of the author. But it is also a convenient way to construct problems that are multilayered, staggered in their complexity: a single-step question to test the basics, a two-step problem to induce the student to think about connecting two different processes, and a multi-step or an open-ended problem that requires one to combine several approaches or use creative thinking. When planning experiments, or making initial assessments of a geochemical system, one often needs to know only an approximate, rather than exact, value for the quantity of interest. Practitioners often acquire the necessary feeling for such numbers through experience, but a skill of making quick estimates is particularly important for students who only begin learning the discipline. A default skill for students in physics and engineering, it is often under-developed in many students in Earth and Environmental Sciences. Perhaps more than other textbooks on geochemistry, this book focuses on employing such estimation techniques.

Finally, a set of Appendices at the end of the book compiles some essential reference information, which is referenced throughout this book.

1.3   A Necessary Word of Caution

This being a short course, many of the details had to be omitted from this book. Occasionally (rarely, I hope), simplifications had to be made for the sake of presenting a clearer story. The world is messy, and a good conceptual model makes it easier to understand. Sometimes, alas, this happens at the expense of occasional failures. Neglecting air resistance helps us understand the laws of projectile motion but fails to describe parachutes or Frisbees. Much in the same fashion, considering geochemical processes in pure water at pH 7 or making some other simplifying assumptions can help us understand important trends, while occasionally failing to account for some types of observations in specific environments. The reader, of course, would be well advised to seek a better explanation for such cases in other monographs.

Chapter 2

Physical Processes

The physical structure of the sediment is the stage on which the action of all biological and chemical processes unfolds. We begin by describing the physical environment within the sediment and the processes that transport its solid and dissolved constituents. These include the burial of deposited layers by newly accumulating sediment, diffusion of molecules in the interstitial water, and the effects of sediment-dwelling organisms. By redistributing the reactants and products of biogeochemical reactions, these processes regulate reaction rates and shape the distributions of chemical species. This chapter introduces formulations for the physical fluxes that move chemicals within sediments and reviews their typical spatial and temporal scales.

2.1 The Physical Structure of Sediments

Sediments are mixtures of solid components and the water that fills the voids. The properties of the solid matrix, as well as of the pore space that it defines, depend on the nature of the settled material. A standard way of classifying the mechanical properties of the sediment is by the particle size of its dominant fraction. *Clay* is composed of the smallest particles (up to 2 micron (μm) in size). Slightly larger particles (2–62 μm) produce *silt*. Fine-grained sediments that may contain a mixture of clay and silt may be referred to as *muddy*. Small particles tend to have large specific surface areas and give the sediment cohesion. They hold water well, and resist hydrodynamic

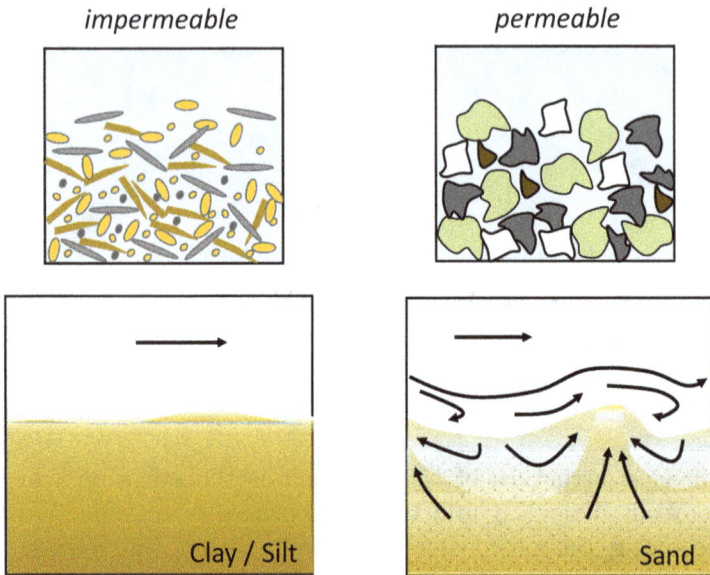

Fig. 2.1. Structure and permeability of sediments. Arrows indicate water flow.

flows. *Sandy* sediments are composed of coarser particles (62 μm to 2 mm). They tend to have high mechanical strength but allow water to flow through them, making them *permeable* (Fig. 2.1). In productive lakes, sediments also can contain large amounts of organic matter, such as decaying remains of aquatic plants and animals and the soil material from the watershed. These organic-rich sediments tend to be loosely bound, having poorly defined particle sizes, and have high amounts of flocculent material at their surface.

The fraction (or the percentage) of sediment volume occupied by water is termed *porosity*:

$$\phi = \frac{V_{\text{porewater}}}{V_{\text{sediment}}} \tag{2.1}$$

Although it is formally a dimensionless quantity, it has an implied unit that is convenient to keep in mind when doing unit conversions: cm^3 of porewater per cm^3 of bulk sediment ($\text{cm}^3_{\text{pw}}/\text{cm}^3_{\text{sed}}$). Recently deposited sediments tend to have porosities on the order of 90%, with the topmost layers containing as much as 99% water by volume. Deeper, more consolidated, sediments have lower porosities, often having ϕ values around 80–85%. The volume fraction occupied by

the solid fraction is

$$\phi_s = 1 - \phi \tag{2.2}$$

The implied unit for ϕ_s is cm^3_{dw}/cm^3_{sed}; the subscript "dw" here stands for *dry weight*, which refers to the solid fraction remaining after drying the sediment.

Test your understanding: Figure 2.2 illustrates the porosity profile (i.e. changes in porosity with depth below the sediment surface) in the sediment of Lake Superior. Suppose you take a 10 cm deep cylindrical sample of this sediment with the cross-sectional area of 100 cm^2. Approximately, how much water could you potentially extract from this sample, for example by placing it in a centrifuge? (The solution is on page 191.)

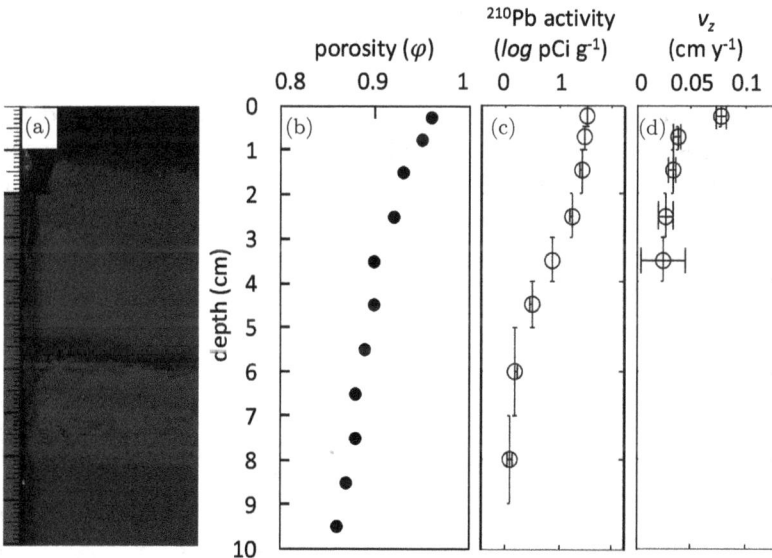

Fig. 2.2. Porosity and compaction in Lake Superior sediment. (a) A visual image of the sediment core. (b) Porosity measurements performed on discrete sediment slices. (c) Decrease in the ^{210}Pb radioactivity that was used to determine the ages of sediment layers (see Section 10.1.1 and Appendix E). Those ages, in turn, were used to calculate the changes in the burial velocity v_z. (d) That were caused by sediment compaction (redrawn from Ref. [11]).

2.2   Physical Transport

2.2.1   Sedimentation, compaction, and burial

Sediments accumulate because solid particles are constantly settling to their surfaces from the overlying water. This downward flux of particles is usually expressed as the mass of solid material deposited per unit area per unit time $(\mathrm{g\,cm^{-2}\,y^{-1}})$. In this formulation, it is known as the *mass accumulation rate* (MAR). The thickness of the sediment that accumulates each year, however, depends on how compactly these particle become arranged within the sediment. In freshly deposited sediments, the solid fraction occupies only a small fraction of the volume: ϕ_s is typically on the order of 1–10%. The accumulation velocity $(\mathrm{cm\,y^{-1}})$ of the freshly deposited sediment is then

$$v_0 = \frac{\mathrm{MAR}}{\rho_s \phi_s^0} = \frac{\mathrm{MAR}}{\rho_s(1 - \phi^0)} \tag{2.3}$$

where ϕ^0 is the porosity near the sediment surface, and ρ_s is the density of solid fraction. For typical mineralogies, ρ_s is around 2.50–2.65 $\mathrm{g/cm^3_{dw}}$.

The porosity decreases in deeper sediment layers, as sediment compresses under its own weight. This *compaction* decreases the thickness of annually deposited layers as they become buried deeper (Fig. 2.3). If one defines the *burial velocity* (cm/y) as the thickness of sediment that accumulated per unit time, it can be defined at each

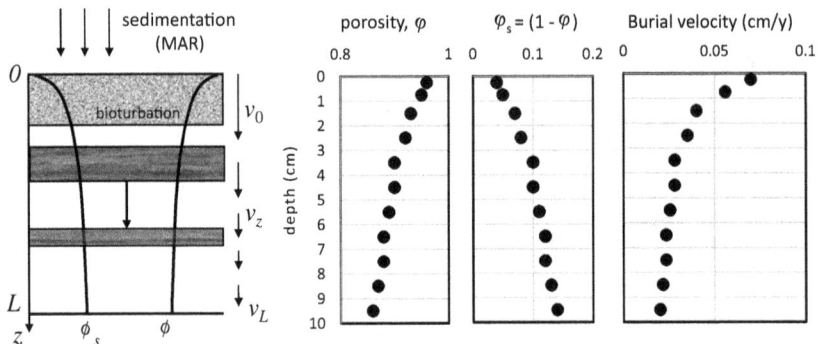

Fig. 2.3. Compaction and sediment transport by burial. The graphs illustrate the effect of compaction on porosity and the resultant change in the burial velocity according to Eq. (2.5).

depth z within the sediment as

$$v_z = \frac{\text{MAR}}{\rho_s \phi_s(z)} = v_0 \frac{\phi_s^0}{\phi_s(z)} \qquad (2.4)$$

Confusingly, the MAR, v_z, and v_0 are all often referred to as *sedimentation rate*. To distinguish among them, one needs to pay close attention to context and units. For practical purposes, saying that the sedimentation rate is 1 cm per year usually implies that the top 1 cm of sediment was deposited within the last year. Because of compaction, however, the material deposited in the first month of that year would occupy a smaller depth interval within the sediment than the topmost material from the last month. Note that a small change in porosity ϕ may mean a large change in the solid fraction ϕ_s, with a correspondingly large change in burial velocity (Fig. 2.3). For example, a decrease in porosity from 95% to 90% means that the fraction of the solid material in each cubic centimeter of the sediment doubles, increasing from 5% to 10%. The speed with which this material becomes buried then halves. As an alternative to the poorly defined burial velocity near the interface, one may use as a reference the velocity v_L that is achieved deep enough inside the sediment $(z = L)$, where compaction becomes negligible:

$$v_z = v_L \frac{\phi_s(L)}{\phi_s(z)} \qquad (2.5)$$

Depending on the practical purpose, this may be often assumed to happen several centimeters or several tens of centimeters below the interface.

Test your understanding: Figure 2.2 illustrates the porosity profile (i.e. changes in porosity with depth) in the sediment of Lake Superior. Approximately, how much solid material (in g/cm^2) is contained in the top 0.5 cm-thick layer of sediment (the porosity there is 0.96)? If the sediment is accumulating at a mass accumulation rate of 0.014 $g\,cm^{-2}y^{-1}$, how many years does it take to deposit this layer of sediment? What will be the thickness of this layer when it is buried down to the depth of 4 cm below the sediment surface? (The answer is on page 191.)

The pile of sediment at the bottoms of lakes and oceans continually grows above the bedrock. For geochemical purposes, however, the most important distance is the depth below the sediment-water interface. It is therefore convenient to fix the zero of the depth scale $z = 0$ at the interface, and count the positive values of z downward (Fig. 2.3). In this frame of reference, as sediment layers become buried, they move downward from the interface, along the z-axis. The burial velocity v_z then represents the speed of the downward burial motion[1] relative to the sediment surface. This frame of reference will be used throughout this book.

The downward burial motion transports the material into the deeper sediment. The corresponding flux $(\mathrm{mol\,cm^{-2}\,y^{-1}})$ of a solid component whose concentration in sediment is C $(\mathrm{mol\,cm^{-3}_{sed}})$ is then

$$F = v_z C \qquad (2.6)$$

This *burial flux* (Fig. 2.4) is responsible for the ultimate removal of reactive substances, such as organic matter, from the ecosystem. As concentrations of solid substances are usually expressed as amounts per gram of solid sediment, rather than per cubic centimeter of wet sediment, the above equation can be rewritten as

$$F = v_z W \rho_s \phi_s \qquad (2.7)$$

where W is the concentration in $\mathrm{mol/g_{dw}}$. As can be seen from Eq. (2.4), when sedimentation is steady and the substance is not being consumed or produced by chemical reactions, the flux remains the same at all depths (matter is conserved) and equal to MAR. Also, the quantity $v_z \phi_s(z)$ remains constant.

Determining sedimentation rates: Where annually deposited layers are easily visible in the sediment vertical sections, sedimentation (burial) velocities can be calculated simply by counting the number of such annual *varves* per unit length. Mass accumulation rate can be then estimated from Eq. (2.3), if the porosity is also estimated or measured. Annual varves, however, are rarely visible in oxygenated environments where sediments are typically mixed by benthic organisms. This usually necessitates the use of sedimentation tracers. A commonly used method relies on the natural decay

---

[1]This is the speed of the solid sediment fraction. As interstitial water is gradually a squeezed from the compacting sediment, it is buried a little slower. For detailed treatment, see Ref. [12].

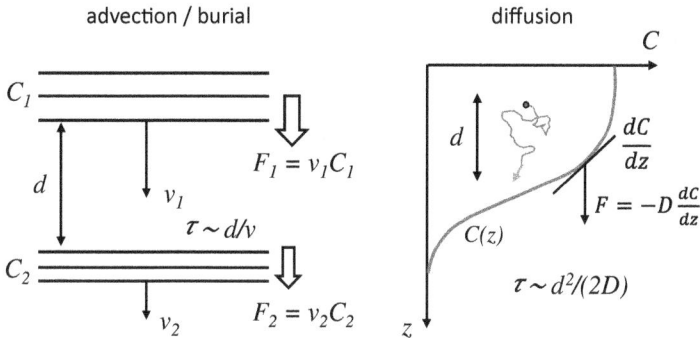

Fig. 2.4. Examples of parameter estimation for physical processes. Fluxes F are estimated from the characteristic transport parameters (v, D) and the concentrations C. The characteristic time scales τ are obtained from distances d and the respective transport parameters.

of the radioactive isotope of lead ^{210}Pb. This isotope is produced in the atmosphere from the radon gas (^{222}Rn) and then transported to the ground by precipitation and further to sediments in association with silicate particles. As ^{210}Pb naturally decays with a half life of 22.3 years, the decrease in the ^{210}Pb activity with sediment depth can be used to calculate the sedimentation velocity (Fig. 2.2) for about the last century. As the sediment surface may be continually homogenized by benthic animals, the width of the often-seen plateau at the top of the sediment ^{210}Pb profile can be used as an indicator of the biological mixing depth. In special cases, approximate sediment *dating* may be also performed by locating the sedimentary layers whose timing is already known, such as those corresponding to some known changes in the lake environment, volcanic tephra deposits, or radioactive ^{137}Cs fallout from the Chernobyl incident in European lakes.

2.2.2 Molecular diffusion

As all molecules in fluid are in constant thermal motion, each single molecule continually experiences random jolts, as it collides with other molecules. As a result, it moves randomly in a *Brownian motion*. When applied to a tracer molecule in a dilute solution, this

Table 2.1. Bulk molecular diffusion coefficients D_0 (cm^2 y^{-1}). Values are calculated based on Refs. [2,13]. Uncertainties can be as large as 20%. More details on the calculation and values for additional species can be found in Ref. [2].

Species	D_0 4°C	25°C	Species	D_0 4°C	25°C
Neutral			Anions		
H_2	1290	1387	OH^-	955	1679
O_2	671	722	Br^-	371	663
CO_2	563	605	HCO_3^-	194	376
H_2S	616	662	$H_2PO_4^-$	155	303
CH_4	490	527	HS^-	362	543
NH_3	668	719	NO_2^-	367	586
Cations			NO_3^-	349	605
H^+	1912	2942	$Acetate^-$	182	345
NH_4^+	352	625	CO_3^{2-}	162	239
Ca^{2+}	136	255	HPO_4^{2-}	125	242
Fe^{2+}	123	223	SO_4^{2-}	183	337
Mn^{2+}	120	222	PO_4^{3-}	101	195

motion causes *molecular diffusion* of the tracer (Fig. 2.4). The mean[2] distance traveled by the tracer molecules increases with time and is given by the famous Einstein's formula for Brownian motion[3]

$$\bar{x} = \sqrt{2Dt} \qquad (2.8)$$

Here, t is time and D is the molecular *diffusion coefficient*. This coefficient characterizes the speed of diffusion and depends on the viscosity of the fluid, the size of the molecule and how strongly it interacts with the molecules of the solvent, as well as on temperature. Larger molecules move slower than smaller molecules, and charged molecules (ions) move slower than neutral molecules (Table 2.1).

[2]Root mean square, strictly speaking.
[3]In his 1905 paper, Einstein explained the Brownian motion by applying a molecular theory of heat to fluids, at a time when the nature of atoms and molecules was still a subject of intense debate.

The fact that the characteristic distance of molecular diffusion (Eq. (2.8)) scales as the square root of time has important implications. Let's invert Eq. (2.8) to express the characteristic time scale of diffusion:

$$t = \frac{\bar{x}^2}{2D} \qquad (2.9)$$

For small distances comparable to the size of a single bacterium ($\bar{x} \approx 1$ μm), diffusion with a typical diffusion coefficient $D = 100$ cm^2/y takes only a millisecond! As water flows at these small spatial scales are dampened by viscosity, microbes exchange substances with their surroundings primarily by molecular diffusion. The diffusion times, however, increase dramatically with distance. Diffusion over cm-scale distances takes hours, and diffusing over a distance of 10 m would take several millennia. The last result means, for example, that when sediment interstitial fluids become buried below the seafloor by several tens of meters, they become isolated from the surface waters for geologically significant times (Table 2.2).

Test your understanding: Suppose you introduce a bromide (Br$^-$) tracer into the water overlying your sediment sample. Approximately, how many hours would you need to wait for the tracer to reach the depth of 2 cm? Assume that within the sediment the tracer is transported only by molecular diffusion and the sediment is highly porous.

Fick's law of diffusion: When the diffusing substance is distributed in space unevenly, the net effect of random diffusion is to distribute the substance from higher to lower concentrations (Fig. 2.4). The flux is proportional to the difference in concentrations, i.e., the concentration gradient. Mathematically, the gradient is defined as the change in concentration per unit distance. As this definition implies

Table 2.2. Distances and time scales of molecular diffusion (from Eq. (2.9), for $D = 100$ cm^2 y^{-1}).

Distance	1 μm	1 mm	1 cm	1 m	10 m
Diffusion time	1.6 ms	26 min	44 h	50 y	5000 y

that the gradient is negative in the direction in which the concentration decreases, the flux has the opposite sign. In one dimension (e.g., along the vertical z-axis), the flux $(\mathrm{mol\,cm^{-2}\,y^{-1}})$ of substance being transported by diffusion is given by the *Fick's law*:

$$F = -D\frac{\partial C}{\partial z} \tag{2.10}$$

Here, C is the concentration (e.g., mol/L), and we use the partial derivative $\partial/\partial z$ because the concentration may be a function of not only depth z but also time t (and in more complicated cases can also vary horizontally, with x and y).

The diffusion coefficients in Table 2.1 describe the molecular diffusion in bulk medium, without the sediment matrix. In reality, diffusion in sediment porewater is slower because the diffusing molecules must follow tortuous paths around the grains. This lengthening of the diffusive path within the sediment medium is described by a dimensionless parameter called *tortuousity*, θ. It characterizes the degree to which the actual diffusion coefficient D differs from the one in bulk liquid D_0:

$$D = D_0/\theta^2 \tag{2.11}$$

Determining the value of θ in each case is a complicated task, but several parameterizations have been suggested as functions of sediment porosity (see Ref. [2]). A popular one for fine-grained sediments, for example, is known under the name of Archie's law:

$$\theta^2 = \phi^{-1.14} \tag{2.12}$$

although another phenomenological relationship

$$\theta^2 = 1 - \ln \phi^2 \tag{2.13}$$

has been also suggested to provide a reasonable fit to the available data [2].

So far we talked only about the diffusion in liquid phase. For solid particles, the Brownian motion effected by molecular diffusion is too slow to be important. The next section, however, discusses a mixing process whose effect on solid sediment phases is similar to the one of diffusion.

2.3 Bioturbation and Bioirrigation

As long as oxygen is available, the top of the sediment serves as a habitat for bottom-dwelling animals (Fig. 2.5). This benthic macrofauna moves around, displacing sediment particles, and ventilates the water around them. Mixing of particles is called *bioturbation*. The enhanced exchange of solutes is termed *bioirrigation*.

The type of biological activity that predominates depends on the organisms present. Oligochaete and tubificid worms are bioturbators while chironomid larvae both bioturbate and bioirrigate. Marine communities are generally much more diverse than freshwater ones. They feature larger animals, even with the same amount of organic matter available for feeding, and they bioturbate substantially deeper. As a result, Marine sediments are routinely bioturbated to depths exceeding 20 cm. Freshwater sediments usually feature smaller animals

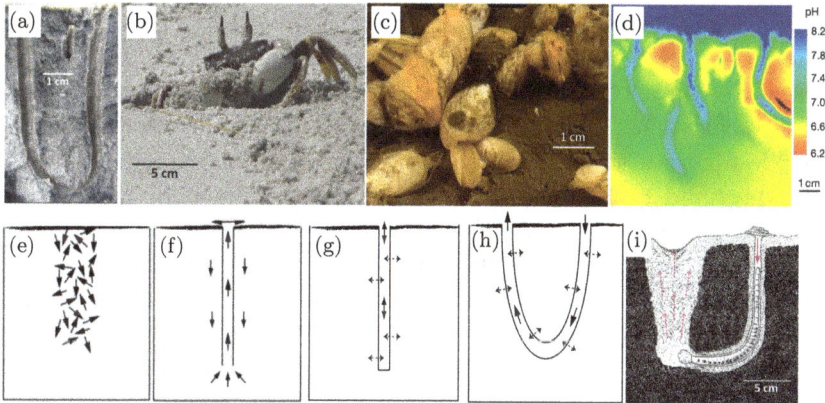

Fig. 2.5. Bioturbating and bioirrigating fauna rework the sediment components in a variety of way, creating complicating three-dimensional patterns in the upper centimeters of sediment. (a) Gallery biodiffusor *Nereis diversicolor* from Denmark; (b) Ghost crab from Thailand; (c) Quagga mussel *Dreissena bugensis* from Lake Michigan; (d) Burrow structure and their ventilation visualized experimentally via two-dimensional pH patterns; (e, f): Examples of particle reworking by bioturbators: (e) biodiffusors; (f) upward conveyors. (g, h) Examples of burrow ventilation patterns by bioirrigators: (g) blind-ended; (h) open-ended; (i) Burrow ventilation by *Arenicola marina*.

Source: (c) this author; (a, b, e–i) reproduced from Ref. [14] with permission; (d) reproduced from Ref. [15] with permission from Elsevier.

and the typical depths of bioturbation there rarely exceed a few centimeters.

Each type of bioturbating fauna displace particles in their own way. Some animals nudge the sediment grains aside as they crawl around. Others ingest sediment and pass it through their bodies. Some purposefully move the material vertically, while others move it in all directions. Yet, despite the diverse and clearly non-random way in which sediment particles are transported by benthic animals, the net effect of bioturbation is remarkably similar to diffusive mixing. In effect, over the time scales much longer than the interval between successive displacements of a particle by a passing animal, the motion of each sediment particle resembles the randomized trajectories of molecules in thermal motion [16]. Accordingly, the "solid diffusion" of sediment particles is described by the equivalent of the Fick's law:

$$F = -D_b \frac{\partial C}{\partial z} \qquad (2.14)$$

where C is the concentration of solid component (e.g., in $\mathrm{mol\,cm_{sed}^{-3}}$). The diffusion coefficient D_b is called the *coefficient of bioturbation*. Its typical values (0.1–2 $\mathrm{cm^2\,y^{-1}}$) are about two orders of magnitude lower than the coefficients of molecular diffusion. Nevertheless, over days and weeks, bioturbation effectively homogenizes the upper millimeters to centimeters of sediment.

The intensity of bioturbation decreases with depth. In some cases, where high-resolution data for the distributions of dissolved oxygen and particulate organic carbon is available, the functional dependence $D_b(z)$ may be calculated from oxygen profiles (see Ref. [11]). Otherwise, more commonly, the intensity of bioturbation is typically measured in incubation experiments, by observing the temporal evolution of non-reactive tracer particles, such as microscopic fluorescent beads [17].

2.4   Fluxes Across the Sediment-Water Interface

One of the main reasons for studying processes inside sediments is to determine the rates at which the sediment exchanges substances with overlying water. Fluxes of nutrients across the sediment-water interface can strongly impact the biological productivity in the water column, and sediment consumption of oxygen can deplete oxygen from the water column.

As discussed above, dissolved substances are exchanged by molecular diffusion and bioirrigation, and their fluxes are driven by concentration gradients between the topmost layer of the sediment and the overlying water. Molecular diffusion fluxes may be estimated from measured concentration profiles (Figs. 2.4 and 2.6). As the molecular diffusion coefficients are known (Table 2.1), diffusive fluxes

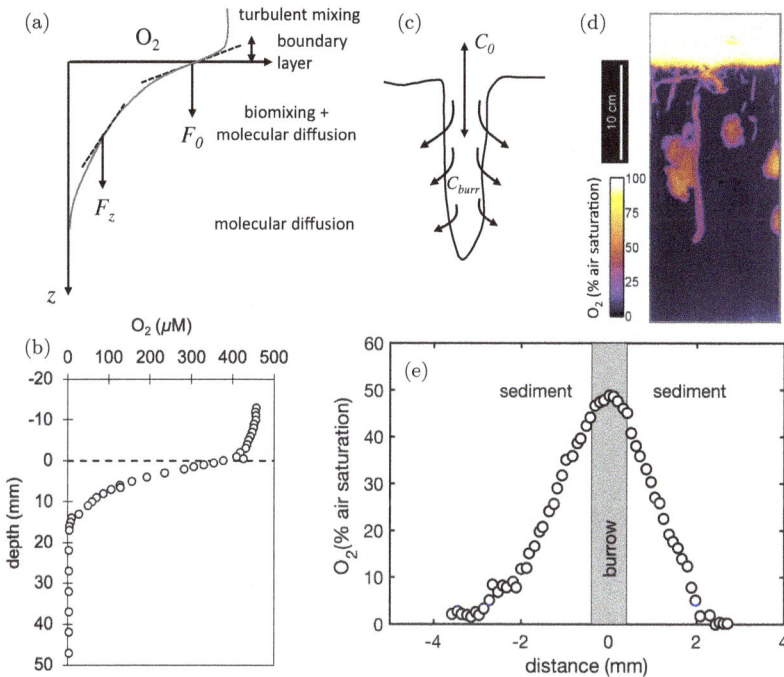

Fig. 2.6. Distributions of solute concentrations across the benthic boundary layer and around animal burrows. (a) Typical shape of the concentration profile of dissolved oxygen. Concentration begins to decrease from its bottom-water value at the top of the benthic boundary layer, a few mm above the sediment-water interface. (b) The one-dimensional profile of dissolved oxygen obtained with a microsensor in the sediment of Lake Michigan. (c) Animal burrows enhance the transport of solutes between overlying water and sediment. (d) Planar optode imaging of oxygen distribution around the polychaete *Clymenella torquata* bioturbated burrows (e) Lateral distribution of oxygen measured across an animal burrow in a coastal marine sediment.
Source: Reproduced from Ref. [18] with permission from Elsevier; (e) Reproduced from Ref. [19] with permission.

are given by Eq. (2.10), with the diffusion coefficient appropriately corrected for temperature and turtuosity. The concentration gradient can often be approximated from discrete measurements in the upper sediment layer. This concentration gradient varies continuously across the *benthic boundary layer*, but is usually steepest near the sediment-water interface. The water above the sediment is in turbulent motion, and transport rates far exceed those of molecular diffusion.

Thus estimated molecular diffusion flux, however, accounts for only a fraction of the total flux. Bioirrigation commonly accounts for up to 50% or more of the total transport of solutes [20]. In sediments where bioturbating fauna are active, total fluxes thus may need to be determined by other methods, such as whole-sediment incubations. The situation becomes even more complicated in permeable sediments, where hydrodynamic flows and pressure differences within the sediment may affect the vertical distributions.

Because in the upper sediment layer bioturbation usually dominates over burial, fluxes of solid substances, such as particulate organic carbon (POC), generally cannot be found from their measured gradients near the sediment-water interface.

2.5   Quick Estimation of Rates, Fluxes, and Time Scales

Equations presented in this chapter enable a number of useful quick calculations for some of the important parameters of diagenesis. For instance, when bioturbation is minimal or shallow, the time horizon for a sediment layer at depth L can be estimated roughly as $t = L/v$, where v is the burial velocity (Fig. 2.4). A more precise estimate that takes into account the sediment compaction may be obtained by integration using Eq. (2.4). For diffusive processes (both molecular diffusion and bioturbation), the characteristic time or depth scales can be calculated from Eqs. (2.8) and (2.9). In muddy sediments, the fluxes of solutes can be approximately estimated from measured concentration profiles using Eq. (2.10) (Fig. 2.4).

These calculations can be done with minimal data and can be used in planning sampling strategies or verifying the consistency of the obtained experimental data. Some of the applications are explored in the following exercises.

2.6 Exercises

1. The sedimentation rate is given by a mass accumulation rate of 0.1 g cm^{-2} y^{-1}. Porosity is 0.9. How many centimeters of sediment accumulate in 100 years?

2. Dissolved bromide (Br$^-$) is used as a non-reactive tracer to investigate the rate of solute transport within sediment. At $t = 0$, bromide is introduced in the water overlying the sediment at a concentration of 1 mM. In the absence of bioturbation and bioirrigation, how deep into the sediment would you expect the bromide to diffuse in 1 day?

3. A non-reactive solid tracer (glass microbeads) is introduced to the surface of the sediment. The bioturbation coefficient is $D_b = 2$ cm^2/y and is approximately constant within the upper layer of sediment. To what depth would you expect the tracer to be mixed by the benthic macrofauna in 1 day?

4. The concentration of oxygen in sediment porewater varies approximately linearly from 100 μM at the sediment-water interface to zero at a depth of 0.2 cm below the interface. What is the flux of oxygen into the sediment? Assume a constant porosity of 0.9.

5. A sediment core is recovered from the bottom of the ocean in a cylindrical tube 10 cm in diameter (see e.g., Appendix A). There is 1 L of overlying water above the surface of the sediment. The core is immediately capped with no air bubble above the water. Researchers need to conduct an experiment for which it is critical that the oxygen concentration in the overlying water does not drop by more than 20% from its initial value. How much time do they have? The sediment is known to have the total oxygen demand of 20 mmolO$_2$ m^{-2}d^{-1}; the initial oxygen concentration in the overlying water is 200 μM.

Chapter 3

Chemical Processes

This chapter deals with chemical reactions that take place in sediments. As described in more detail in the next chapter, many important chemical transformations in sediments are catalyzed by microorganisms, so the rates of reactions are intimately dependent on microbial activity. In this chapter, we focus on the non-biological fundamentals that regulate the chemical aspects of sediment diagenesis. We define the necessary basic concepts and review the thermodynamic and kinetic factors that govern the dynamics of chemical reactions and the precipitation and dissolution of minerals. The last section of the chapter examines the combined effects of chemical reactions and physical transport and analyzes how they shape the vertical distributions (profiles) of substances within sediments.

3.1 Chemical Processes in Sediments

Geochemistry is built around the *cycling* of elements—that is, the networks of processes that move the atoms of chemical elements physically through the environment and chemically between different types of molecules, creating and destroying compounds. These processes regenerate and recycle substances within the system and exchange them with the outside environment. The most important chemical cycles in sediments are those of the essential elements of Life: carbon, macronutrients (nitrogen and phosphorus), and the associated elements iron, manganese, and sulfur. This book devotes separate chapters to each of these cycles, describing the corresponding chemical transformations and the reaction networks in more detail. This chapter focuses, instead, on the more general factors that control the outcomes of geochemical transformations, as well as the rates at which such transformations occur.

The favorability of reactions, defined largely by whether energy is released or consumed, is addressed using the concepts of thermodynamics. A particularly important class of reactions are redox reactions, in which electrons are exchanged between atoms. Electrons are the currency of chemical cycles: The direction in which they flow underlies the classification of environments as oxidizing or reducing. Such reactions are addressed here in some detail. Addressing the speed of reactions requires concepts from chemical kinetics. A convenient shortcut is to distinguish between very fast reactions where the reactants and products come into equilibrium very quickly, and those in which the rate of transformation is slow enough so that a

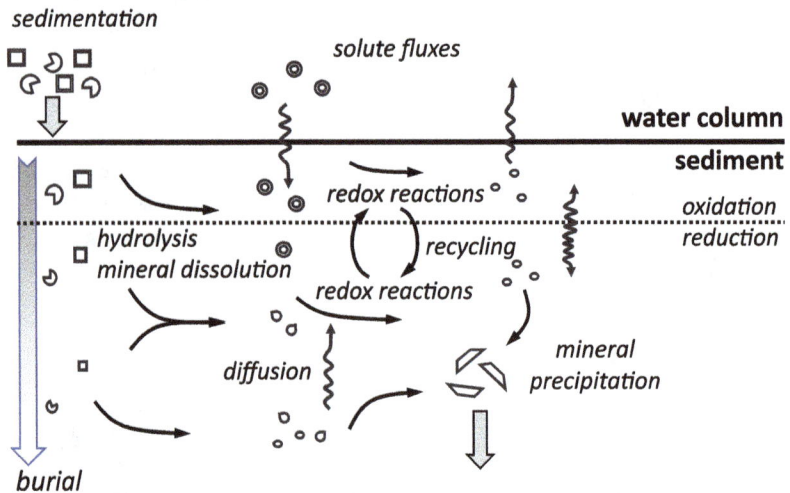

Fig. 3.1. Illustration of the processes involved in elemental chemical cycling in sediments. Chemical elements are transported by burial and diffusion. Solid organic compounds can become hydrolyzed and products mobilized into the porewater. Deposited minerals can dissolve, while new (authigenic) minerals can form by precipitation within the sediment. Multiple redox (reduction and oxidation) reactions transform the chemicals within the sediment. Repeated redox transformations, coupled with the transport of reactants and products across the sediment redox boundary, can recycle elements within sediments multiple times. They also determine the fluxes of substances to and from the ambient environment, and depend on environmental variables such as the pH.

degree of disequilibrium is constantly maintained. We may thus refer to them as, respectively, *equilibrium reactions* and *kinetic reactions*. In addition to chemical reactions *per se*, which involve creating or breaking chemical bonds between atoms, this chapter also addresses the processes of mineral precipitation and dissolution. Even when redox reactions (e.g., between two soluble compounds) may lead to the formation of an insoluble compound that would precipitate out of solution, mineral precipitation itself does not involve a transfer of electrons and should be considered as a separate type of process, with its own mechanisms and control factors.

3.2 The Basics

3.2.1 Concentrations, fluxes, and rates

We begin with a primer on terminology. What should be called a "rate", and how is it different from a "flux"? Despite the question seeming to be trivial, different branches of science use slightly different definitions for these terms, which generates much confusion. While not trying to impose a single pedantic formulation, this book uses the convention described below. The key is to pay attention to the dimension (units) of the quantity of interest.

Amount (Q) is the quantity of substance, e.g., in kg or mol.

Concentration (C) is the amount of substance per unit volume: $C = Q/V$. Accordingly, the unit for concentration is kg/m^3 or mol/m^3. Whereas liter (L) is a commonly used unit for volume, there are good reasons for using length cubed [21], especially where unit conversions are involved. The conversion is $1\,L = 1\,dm^3 = 10^{-3}\,m^3 = 10^3\,cm^3$.

Suppose our study system is a box that contains a solution of some chemical substance at concentration C (Fig. 3.2). This concentration may change if either: a) the substance is added to (or removed from) the box; or b) the substance is produced (or consumed) within the box. The first case reflects the effect of *fluxes* across the boundaries of the system. The second reflects the effect of processes within the system, whose fastness is typically described as *rates*. Thus,

Flux (F) is the amount of substance transported across a boundary of the system per unit area per unit time. Fluxes therefore may be measured in units such as $mol/m^2/s$. For example, if a river flows

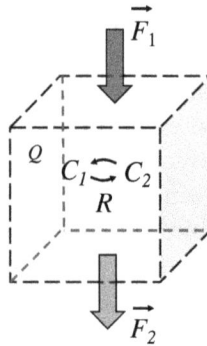

Fig. 3.2.   Diagram: Fluxes across the system boundaries and rates within the system.

into a lake and the speed of the current is $v = 2$ m/s, and the concentration of a certain chemical in the river is $C = 3$ mmol/m^3, then the flux of that chemical into the lake is

$$F = Cv \quad (\text{mmol/m}^3 \times \text{m/s} = \text{mmol/m}^2/\text{s}),$$

or 6 mmol m^{-2} s^{-1}.

Rates (R) generally characterize how quickly a certain process occurs within the system. As such, their definitions typically involve a derivative with respect to time. Chemical reaction rates, in particular, can be defined as the change in the amount of substance per unit volume per unit time, or the change in concentration per unit time: $R = dC/dt$. The unit in this case is mol m^{-3} s^{-1}.

Note that, for these definitions, rates are generally associated with changes over time, whereas fluxes are associated with movement through space or across boundaries. Another way to express this is to notice that fluxes are essentially vectors whose directions point in the direction of the flux. A reader familiar with physics may notice a similarity here to the definitions of quantities such as magnetic flux or the flux of electric field, which are defined as vectors. Rates, on the other hand, are scalars. In hydrology, however, terms like "flow rate", are commonly used for water flows that are expressed volumetrically as m^3/s (cubic meters of water flowing through a river cross-section per second). This is essentially a flux of water, as its meaning is the number of cubic meters of water that flow through each m^2 of the cross-section per second, multiplied by the cross-sectional area. Likewise, the inputs and outflows of chemicals into or from a lake can be expressed as amounts added per second (mol/s), in which case they are often called "input rates". Despite the word "rate", these are still measures of fluxes across system boundaries. Whatever the term, keeping track of the units will help one keep track of the meaning.

A note on units: Although the International System of Units (SI) has been introduced and widely adopted long ago, consistent use of standardized units has not fully percolated into some fields. Accordingly, you may find the concentrations of dissolved oxygen in water to be reported in mol/L, mg/L, or ml/L, and seemingly never in the SI units of mol/m^3. While the choice of grams (g) vs kilograms (kg) is largely a matter of convenience and does not cause much trouble, the choice of molar vs. mass units can lead to confusion. Comparing 2 mg/L of nitrogen gas (N_2) to 5 mg/L of nitrate (NO_3^-), for instance, does not give a clear idea in which case the concentration of nitrogen is greater. And the author is painfully aware of situations where confusion about whether to use the molar mass of 16 (for atomic oxygen) or 32 (for gaseous O_2) when dealing with oxygen concentrations in mg/L can lead to a factor of two difference in the calculated concentrations and some embarrassing gaffes in press releases. For reasons such as these, the use of the molar units (e.g. mol/L) is advised (as is widely adopted in oceanography).

Test your understanding: Within a certain sediment layer, diagenetic reactions produce ammonium (NH_4^+) at a rate of 10 μmol/L/h. The ammonium diffuses away from that layer, so the concentration within the layer remains constant. The layer is 1 cm thick. What is the flux of ammonium from the layer? (The answer is on page 192.)

3.2.2 Activities and fugacities

We will use square brackets to indicate **concentrations** of chemical species. For example, $[HCO_3-]$ would indicate the concentration of bicarbonate ions.[1] The standard unit of concentration is molar (M),

---

[1]Chemical literature sometimes reserves square brackets for equilibrium concentrations, whereas ambient concentrations are written using curly brackets. We will use one type of brackets here, for simplicity, and will mark equilibrium concentrations explicitly where needed.

or mol per liter. One mol of substance contains the Avogadro number $(N_A = 6.022 \times 10^{23})$ of molecules. Since molecules in solutions interact with other molecules, their chemical activity becomes affected by their surroundings and the concentrations of other chemical species. **Activities**, a, are used to describe the effective concentrations in non-ideal solutions. Their relationship to concentrations C is given by $a = \gamma C$, where γ is the *activity coefficient*. By convention, activities are dimensionless quantities whose values are taken to be unity at the substance's chosen standard state. For instance, activities of pure substances, such as solids or pure water, are equal to 1. For solutes, the standard state is a hypothetical solution with the concentration of 1 mol/L. Thus, in dilute solutions, which may be considered as nearly ideal solutions, the activity coefficient is close to 1 and activities can be taken as being approximately equal to concentrations (though formally having different units).

For gases, the equivalents of concentrations and activities are, respectively, **partial pressure**, p, and **fugacity**, f [22]. In an ideal gas, they are equivalent. Unlike activity, which is dimensionless, fugacity is measured in the same units as pressure. For example, methane presently makes up about 1.8×10^{-6} of the Earth's atmosphere, so its fugacity approximately equals its partial pressure of 1.8×10^{-6} atm.

3.2.3 Solubility of gases

When a gas is in contact with a solution, such as at the surface of the ocean or at the surface of a gas bubble inside the sediment, molecules can cross between phases. At equilibrium, the concentrations of the molecules in the dissolved and gaseous states are linked via the Henry's law:

$$C_{aq} = K_H P \qquad (3.1)$$

where C_{aq} is the concentration of the dissolved gas and P is the partial pressure of the gas in its gaseous phase. Table C.4 lists the values of the Henry's Law constant K_H. At high concentrations or high pressures, C_{aq} and P should be replaced, respectively, by activity and fugacity. The Henry's Law constant K_H is temperature-dependent: more gas can be dissolved in cold than in warm water (Section C.4).

3.2.4 The pH

Like all molecules, water molecules H_2O are in constant thermal motion. This thermal energy causes them to constantly and reversibly dissociate into the H^+ and OH^- ions.

$$H_2O \longleftrightarrow H^+ + OH^- \tag{3.2}$$

These charged species can have strong effects on chemical reactions, and their concentrations can be, in turn, affected by the sediment processes. The **pH** of a solution reflects the activity (concentration) of protons, $[H^+]$, and is defined as

$$pH = -\log_{10} a_{H^+} \approx -\log_{10}[H^+]$$

As the equilibrium concentration of H^+ in pure water is 10^{-7} M (see Section 3.3.2), it corresponds to the pH of 7.0. This value serves as the neutral point between acidic (pH $<$ 7) and basic (pH $>$ 7) solutions. Note that the pH is a logarithmic quantity, so a change in the pH by one unit corresponds to a tenfold change in the H^+ concentration. Given that water molecules actively participate in reactions in aqueous solutions, the pH of the solution can strongly affect the outcomes of reactions.

3.3 Thermodynamics of Chemical Reactions

3.3.1 Gibbs free energy of reactions

The direction of chemical reactions in the environment strongly depends on whether there is energy to be gained. Let us consider a generic reaction:

$$aA + bB \longrightarrow cC + dD \tag{3.3}$$

Here, A and B are reactants, C and D are products, and a, b, c, d are stoichiometric coefficients. If the chemical energy contained in the compounds A and B is greater than the chemical energy in C and D, the reaction will occur with the release of energy, and the net reaction will proceed in the direction indicated by the arrow in Eq. (3.3). That is, while the reverse reaction may still randomly occur, it requires additional energy, so it would occur less frequently than the forward reaction. The quantity that is most relevant for energy changes in chemical transformations is the *Gibbs free energy*, G, which is one of several *thermodynamic potentials*. In accordance with the second law of thermodynamics, in a thermodynamically closed system at constant temperature and pressure, the Gibbs free energy of the system can only decrease (corresponding to a production of entropy) [22]. This decrease in chemical potential energy is commonly talked about as *energy gain*, as the energy becomes available for use in other forms (mechanical work, biomass synthesis, etc.).[2]

The energy gain of a reaction depends on the concentrations of the reactants and products. Values of Gibbs free energies for individual reactants and products are often tabulated (Appendix C) under the conditions chosen as standard: e.g., temperature 25°C, atmospheric pressure, and chemical concentrations of 1 mol/L. These are known as the Gibbs free energies of formation (from the elements). Accordingly, the change in the Gibbs free energy of the reaction, ΔG^0, can be calculated under standard conditions from their differences. These standard energy changes ΔG^0 are also often tabulated for each reaction (Appendix D). Standard conditions, however, rarely exist in natural systems. For dissolved constituents, for example, they assume the unrealistically high concentrations of 1 M, which are virtually never achieved. The energy gain under non-standard concentrations can be calculated as

$$\Delta G = \Delta G^0 + RT \ln Q \tag{3.4}$$

Here, T is the absolute temperature (in degrees Kelvin) and $R = 8.31\,\text{J}/(\text{mol}\,\text{K}) = 8.31 \times 10^{-3}\text{kJ}/(\text{mol}\,\text{K})$ is the universal gas constant. The *reaction quotient*, Q is calculated as the product of the corresponding activities:

$$Q = \prod_i a_i^{\nu_i} \tag{3.5}$$

---

[2]A decrease in the Gibbs free energy G is associated with a decrease in enthalpy H and an increase in entropy S: $\Delta G = \Delta H - T\Delta S$. The enthalpy change corresponds to heat production, while the production of entropy can be associated with breaking of complex molecules into simpler ones (i.e. the generation of disorder). To synthesize complex molecules in microbial cells, for example, the decreased entropy of the cell needs to be offset by a discharge of entropy (as simple metabolite molecules) and enthalpy (heat) into the environment.

Here, $a_i^{\nu_i}$ is the activity of the ith reactant or product, and ν_i is the corresponding stoichiometric coefficient in the reaction (negative for reactants). Using the reaction (3.3) as an example, for dilute solutions when concentrations may be used instead of activities, the reaction quotient then becomes

$$Q \approx \frac{[C]^c[D]^d}{[A]^a[B]^b}$$

The differences between the concentrations and activities may become significant at high concentrations, as well as in saltwater [22] (see Appendix C).

Example: Let us estimate the thermodynamic favorability of the reaction that forms (a rather insoluble) iron monosulfide from dissolved ferrous iron and ionic hydrogen sulfide:

$$Fe^{2+} + HS^- \longrightarrow FeS(s) + H^+$$

This is a reaction that commonly takes place in anoxic sediments and regulates the concentrations of dissolved Fe^{2+} and hydrogen sulfide in the porewater. The resulting iron monosulfide quickly precipitates, so it is written here as solid phase. The generic notation "FeS" may in reality correspond to several different minerals [23]; here we will use mackinawite for the sake of an example. Using the standard Gibbs free energies of formation in Table C.5, the change in the Gibbs free energy under standard conditions is (products minus reactants):

$$\Delta G^0 = (-93.30 + 0) - (-78.87 + 11.97) = -26.40 \, \text{kJ/mol}$$

This negative difference corresponds to the 26.40 kJ/mol of energy gained under standard conditions. We can then estimate the concentrations of Fe^{2+} and HS^- for which the reaction remains favorable. It happens when $\Delta G < 0$, so from Eq. (3.4):

$$Q < e^{-\frac{\Delta G^0}{RT}}$$

As activities for solid phases are equal to 1 by convention, for a dilute solution this can be written as

$$Q = \frac{[H^+]}{[Fe^{2+}][HS^-]} < e^{-\frac{\Delta G^0}{RT}}$$

At the neutral pH of 7, the concentration of hydrogen ion is $[H^+]=10^{-7}$ M. So, for a sediment at a typical temperature of

$T = 4°C = 277\,\text{K}$, the reaction should be favorable when

$$[Fe^{2+}][HS^-] > 10^{-7}e^{\frac{-26.40\times10^3}{8.31\times277}} = 1.0 \times 10^{-12}\,\text{M}^2 = 1.0\,\mu\text{M}^2$$

(The factor 10^3 in the exponent accounts for the conversion from kJ/mol to J/mol.) Thus, the reaction is favorable when the product of concentrations $[Fe^{2+}][HS^-]$ (in units of micromolar) exceeds 1.0. The dissolved iron and HS^- thus may be expected to co-exist in sediment porewater when the concentrations of both species are at sub-micromolar levels, while at higher concentrations formation of solid FeS should be expected.[3]

A note on speciation: As will be discussed below, hydrogen sulfide exists in two forms, HS^- and H_2S, which are approximately equally abundant at neutral pH (see Appendix C). Accordingly, a similar calculation needs to be carried out also for the alternative reaction, such as

$$Fe^{2+} + H_2S \longrightarrow FeS(s) + 2\,H^+$$

We leave it as an exercise to the reader to verify that it leads to a similar constraint on the concentrations of Fe^{2+} and H_2S. The result should not be surprising, as the equilibration reaction between H_2S and HS^-, being at equilibrium, entails no energy gain ΔG of its own (see Section 3.3.2).

A note on the pH effect: The pH can strongly influence the favorability of reactions. Suppose the reaction above, instead of taking place in a pH-neutral solution, happens at the acidic pH of 4.0. At that pH, nearly all hydrogen sulfide is present as H_2S (Fig. C.3). As the reader should be able to verify, the constraint on the concentration product in that case becomes: $[Fe^{2+}][H_2S] > 3.4 \times 10^6\,\mu\text{M}^2$. That is, at this low pH, dissolved iron can co-exist with dissolved sulfide at very high (millimolar) levels, meaning that for typical concentrations in sediments the precipitation of FeS would be inhibited and its dissolution would be favored instead (see also [24]).

Test your understanding: Hydrogen sulfide (H_2S) is readily oxidized by oxygen to sulfate (SO_4^{2-}):

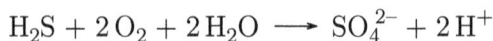

$$H_2S + 2\,O_2 + 2\,H_2O \longrightarrow SO_4^{2-} + 2\,H^+$$

---

[3]A more nuanced discussion, which takes into consideration an important role of the aqueous complex FeS_{aq}, is presented in [24].

Suppose that the limit of oxygen penetration into the sediment is defined as the depth where dissolved oxygen (O_2) is no longer detectable by a probe with the detection limit of 1 μM. The concentration of sulfate at that depth was found to be 100 μM. What is the maximum concentration of hydrogen sulfide that one might expect to find at that depth, based on the thermodynamics of the reaction? (The answer is on page 192).

3.3.2 Equilibrium reactions

While ΔG remains negative, the concentrations of reactants can decrease and the concentrations of products can increase, until the forward and reverse reaction rates become equal and the net reaction produces no energy gain. Such balance corresponds to a state of thermodynamic equilibrium. Some reactions achieve such equilibrium quickly, and are often called *equilibrium reactions*.

The distribution of chemical species at equilibrium is governed by Eq. (3.4) for $\Delta G = 0$, which gives

$$\Delta G^0 = -RT \ln K_{eq} \qquad (3.6)$$

where K_{eq} is the reaction quotient at equilibrium, called the *equilibrium constant*

$$K_{eq} = \prod_i a_i^{\nu_i} \qquad (3.7)$$

The activities a_i here correspond to the equilibrium concentrations of the respective species. For example, for the generic reaction (3.3) (again, assuming a dilute solution so that activities can be represented by concentrations):

$$K_{eq} = \frac{[C]^c[D]^d}{[A]^a[B]^b} \qquad (3.8)$$

Rearranging Eq. (3.6) provides a way of calculating equilibrium constants from the respective ΔG^0 of the reaction:

$$K_{eq} = e^{-\Delta G^0/RT} \qquad (3.9)$$

The values for some important equilibrium constants are given in Appendix C.

Some of the most common (and important) equilibration reactions involve the molecule of water, H_2O. Water molecules continually break down into the H^+ and OH^- ions due to thermal motion, and are being continually reconstituted from these ions (Eq. (3.2)). The association reaction by which the hydrogen (H^+) and hydroxide (OH^-) ions combine to form a water molecule happens in a tiny fraction of a second (while the time scale for the dissociation reaction can be hours [25]). Hence for nearly all processes in sediments it can be considered instantaneous, and the concentrations of its reactants and products achieve the equilibrium defined by the absence of a net energy gain $\Delta G = 0$. Given the standard energies of formation (Appendix C), creation of a water molecule from two ions releases $\Delta G^0 = 79.89$ kJ/mol of free energy. Solving Eq. (3.4) for the concentrations gives an important result: at the standard temperature and pressure, the concentrations of H^+ and OH^- in pure water are each 10^{-7} mol/L. As discussed above, this equilibrium corresponds to the pH of 7, which is defined as neutral.

3.3.3 The carbonate system

An environmentally important example of equilibration reactions is given by a series of reactions that take place when carbon dioxide (CO_2 gas) is dissolved in water. The dissociation of the resulting carbonic acid (H_2CO_3) into dissolved CO_3^{2-} (carbonate ion), HCO_3^- (bicarbonate ion), and non-ionic CO_2 holds a special place in aquatic chemistry. The concentrations of these three chemical species regulate the exchanges of CO_2 with the atmosphere, formation of carbonate minerals, and the pH balance of many natural systems.

The equilibria among these chemical species[4] are governed by fast reactions:

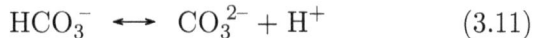

$$CO_2(aq) + H_2O \longleftrightarrow HCO_3^- + H^+ \tag{3.10}$$

$$HCO_3^- \longleftrightarrow CO_3^{2-} + H^+ \tag{3.11}$$

Their equilibrium constants are, respectively,

$$K_1 = \frac{[HCO_3^-][H^+]}{[CO_2]} \tag{3.12}$$

$$K_2 = \frac{[CO_3^{2-}][H^+]}{[HCO_3^-]} \tag{3.13}$$

---

[4]For a closed system. For more complicated cases where the system is open to the atmosphere, see e.g., [22].

Their values are given in Table C.1. The speciation (relative concentrations) of the carbonate species can be calculated from these relationships, if one adds the condition that the total concentration of dissolved CO_2 is the sum of the three dissolved species: $C_{tot} = [CO_2(aq)] + [HCO_3{-}] + [CO_3^{2-}]$. The three equations then can be solved for the three concentration ratios – for each species as a fraction of the total. The result is shown as a function of the pH in Fig. 3.3 (and the analytical solution is listed in Appendix C). It indicates that, at the circumneutral pH, the dominant species is the bicarbonate (around 82%), with aqueous CO_2 making up most of the difference. The CO_2 dominates at acidic pH (below 6.3), whereas the carbonate ion (CO_3^{2-}) becomes important only in alkaline (basic) waters, dominating above pH 10.3.

The carbonate speciation relationship in seawater is broadly similar to this result, but the corresponding calculation requires corrections to account for the activities of other ions, which results in slightly different curves (see e.g., [22]).

Test your understanding: The concentration of the CO_2 gas in the atmosphere is presently 412 ppm (meaning the gas makes up 0.0412% of the atmosphere by volume), which corresponds to the partial pressure of 0.412×10^{-3} atm. Suppose a bubble of air is trapped underwater at shallow depth. What is the concentration of aqueous CO_2 and total dissolved CO_2 (for pure water) that would be in equilibrium with the CO_2 gas inside the bubble? Compare your result to the typical concentrations of dissolved inorganic carbon

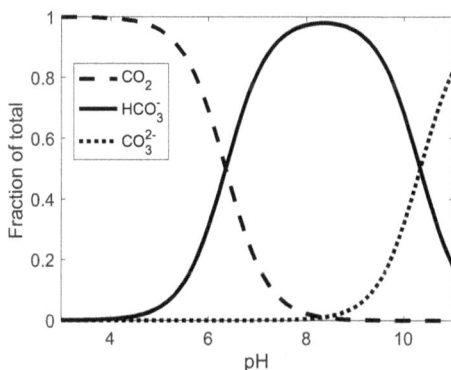

Fig. 3.3. Carbonate equilibrium: Concentrations vs. pH.

(DIC) in lakes and oceans and hypothesize about the typical direction of the CO_2 flux across the water-air interface (you may need to reflect on the difference between ideal and realistic conditions). Compare your results at pH 7 and pH 8 and at 25°C vs. 4°C. (The answer is on page 192.)

3.3.4 Adsorption

Numerous small particles, such as clay minerals, iron oxyhydroxides, and organic matter, in sediments offer abundant surface areas that can adsorb significant quantities of dissolved substances from sediment porewater. For instance, uncompensated electric charges on the surfaces of solid particles may serve as effective adsorption sites for ions of the opposite sign from the solution. The reactions of sorption and desorption are often fast and can be considered to be in local equilibrium. The ways to quantify the adsorbed amounts are described in Section 11.3.5.

3.4 Redox Reactions

An important class of reactions involves a transfer of electrons from one atom to another. The atom that loses the electron is said to be *oxidized*, while the atom that gains the electron is said to be *reduced*. The corresponding reactions that couple the oxidation of one component to the reduction of another are called reduction-oxidation reactions, or *redox* for short. As they include many biologically important reactions, including photosynthesis and respiration, they hold a special place in biogeochemistry, including sediment diagenesis.

3.4.1 Half-reaction representation

The reduction and the oxidation sides of the reaction are often considered separately, in a *half reaction* representation that explicitly shows the transfer of electrons. Such representation is convenient as it allows to combine different reductants and oxidants into *redox pairs*, to form the corresponding overall redox reactions. For example, oxidation of an organic molecule such as glucose can be written independent of a specific oxidant [6]:

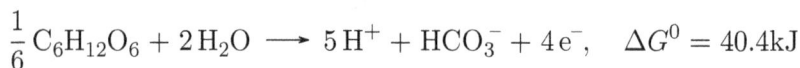

$$\frac{1}{6}\,C_6H_{12}O_6 + 2\,H_2O \longrightarrow 5\,H^+ + HCO_3^- + 4\,e^-, \quad \Delta G^0 = 40.4\text{kJ}$$

The fact that the electrons appear on the product side of the reaction (i.e., the electrons are being donated) indicates that this is an oxidation half-reaction. Similarly, reduction of dissolved oxygen (i.e., accepting the electrons) can be written without specifying the reductant:

$$O_2 + 4\,H^+ + 4\,e^- \longrightarrow 2\,H_2O, \quad \Delta G^0 = -490.68\text{kJ}$$

The two half-reactions can be combined to obtain the overall reaction of the oxidation of glucose by oxygen:

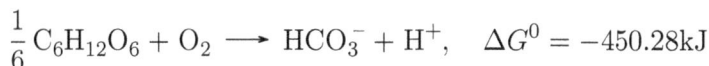

$$\frac{1}{6}\,C_6H_{12}O_6 + O_2 \longrightarrow HCO_3^- + H^+, \quad \Delta G^0 = -450.28\text{kJ}$$

Table 3.1 and Table C.6 list some of the half-reactions that make up the diagenetically important redox transformations.

Test your understanding: Use the half-reactions in Table C.6 to construct the overall reactions for: (a) oxidation of ammonium (NH_4^+) to nitrate (NO_3^-) by oxygen; (b) oxidation of methane (CH_4)

Table 3.1. Selected half-reactions and their redox potentials, calculated from the Gibbs free energies of formation (Table C.5) and Eq. (3.17). See also Table C.6.

Oxidized form	Reduced form	E_H^0 (V)	E_H^0 (V) (pH 7, 4°C)
$O_2 + 4\,H^+ + 4\,e^-$	$2\,H_2O$	1.27	0.89
$NO_3^- + 6\,H^+ + 5\,e^-$	$\frac{1}{2}\,N_2(g) + 3\,H_2O$	1.24	0.78
$MnO_2 + 4\,H^+ + 2\,e^-$	$Mn^{2+} + 2\,H_2O$	1.23	0.46
$Fe(OH)_3 + 3\,H^+ + e^-$	$Fe^{2+} + 3\,H_2O$	0.95	-0.20
$SO_4^{2-} + 10\,H^+ + 8\,e^-$	$H_2S + 4\,H_2O$	0.30	-0.18
$HCO_3^- + 9\,H^+ + 8\,e^-$	$CH_4 + 3\,H_2O$	0.21	-0.22
$HCO_3^- + 5\,H^+ + 4\,e^-$	$\frac{1}{6}\,C_6H_{12}O_6 + 2\,H_2O$	0.20	-0.22
$2\,H^+ + 2\,e^-$	$H_2(g)$	0.00	-0.38

by sulfate (SO_4^{2-}). Calculate the corresponding changes in the Gibbs free energy ΔG^0 for each reaction. If necessary, use the carbonate equilibrium equations to simplify the reactions.

3.4.2   Redox potential and the E_H

The ease with which molecules lose or gain electrons depends on the potential energy of the electrons within them. Accordingly, it is characterized by a quantity E that has the same units as electric potential: Volts. It is called *redox potential*. Similar to the electric potential, which describes the change in the electric potential energy per unit charge, the redox potential describes the change in the potential Gibbs free energy per electron. Like the electric potential, which can be counted relative to some arbitrarily defined level, the redox potential is defined relative to a reference that is chosen by convention. This standard reference level, which is assigned the redox potential of $E^0 = 0.00$, corresponds to the half reaction

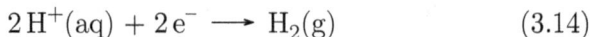

$$2\,H^+(aq) + 2\,e^- \longrightarrow H_2(g) \tag{3.14}$$

at the standard conditions of $a_{H+} = 1$ and $p_{H2} = 1\,atm$. This reduction reaction is known as the *standard hydrogen electrode* (SHE). When measured relative to the SHE, redox potentials (also called "electrode potentials") are designated as E_H. By convention, they are compared for reactions written as reduction reactions. Higher positive values E_H for half-reactions correspond to stronger oxidants (Table C.6). Negative E_H values for the reduction half-reactions indicate potential to reduce.

Like electric potentials in an electric circuit, redox potentials for coupled reactions are additive and do not depend on the number of the electrons transferred. For example, the reaction of oxidation of ferrous iron (Fe^{2+}) by oxygen can be constructed from the corresponding half reactions (see Table C.6):

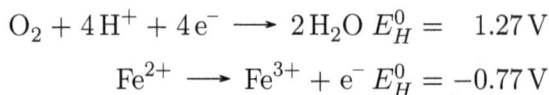

$$O_2 + 4\,H^+ + 4\,e^- \longrightarrow 2\,H_2O \quad E_H^0 = 1.27\,V$$

$$Fe^{2+} \longrightarrow Fe^{3+} + e^- \quad E_H^0 = -0.77\,V$$

$$\overline{}$$

$$O_2 + 4\,Fe^{2+} + 4\,H^+ \longrightarrow 4\,Fe^{3+} + 2\,H_2O \quad E_H^0 = +0.50\,V$$

Note that the stoichiometric factor of 4 for the oxidation half of the reaction does not affect the E_H^0. The positive value of E_H^0 for the overall reaction indicates that the electrons can be transferred with energy gain, i.e., the reaction is energetically favorable.

The thermodynamic favorability of the redox reactions can be expressed through the relationship with the Gibbs free energy (per mol of reaction):

$$\Delta G^0 = -nFE_H^0 \qquad (3.15)$$

where n is the number of electrons transferred in the reaction. The coefficient of proportionality is the Faraday's constant $F = 96.485\,\text{kJ/V/mol}$, which is equivalent to the electrical change of a mol of electrons ($F = eN_A = 96{,}485\,\text{C/mol}$). As Eq. (3.15) indicates, positive values of the E_H correspond to negative values of ΔG^0, i.e., the thermodynamical favorability of the reaction. Unlike the redox potential, the amount of energy gained in the reaction does depend on the number of the transferred electrons n, and thus can change depending on the stoichiometry of the overall reaction.

Note: $p\epsilon$. Geochemical literature often makes use of a parameter $p\epsilon$, which is supposed to be an analogue of the pH for the electrons. It is defined accordingly as $p\epsilon = -\log_{10}(a_{e^-})$, where a_{e^-} is the hypothetical activity of electrons in the solution. (Unlike the protons in the definition of the pH, free electrons do not exist in solution.) The $p\epsilon$ is related to the redox potential as

$$p\epsilon^0 = \frac{F}{2.303RT}E_H^0 \qquad (3.16)$$

Test your understanding: Use the E_H^0 values in Table C.6 to determine whether the following reactions are thermodynamically favorable: (a) oxidation of nitrite (NO_2^-) by oxygen; (b) oxidation of ammonium (NH_4^+) by sulfate (SO_4^{2-}). Calculate the corresponding ΔG^0 values for the overall reactions, to verify your result. Notice that the Gibbs free energy of the reaction depends on the number of transferred electrons, whereas the E_H^0 does not.

3.4.3 Redox potentials under environmental conditions

Just like the Gibbs free energy of the reaction may differ significantly in natural sediments from its value at standard state, redox potentials in natural sediments can differ from their standard E_H^0 values. Combining Eq. (3.4) with Eq. (3.15) gives the corresponding relationship for the redox potential as a function of environmental conditions. It is known as the *Nernst equation*:

$$E_H = E_H^0 - \frac{RT}{nF} \ln Q \qquad (3.17)$$

As the logarithm of the reaction quotient equals the sum of the logarithms of the activities (concentrations) of individual species, it can be used to evaluate the effects of each individual species on the reaction's thermodynamic favorability. For example, for the reaction of sulfate reduction ($E_H^0 = 0.30$ V) in Table 3.1:

$$E_H = E_H^0 - \frac{RT}{8F} \left(\ln [H_2S] - \ln [SO_4^{2-}] - \ln [H^+]^{10} \right) \qquad (3.18)$$

By substituting the appropriate concentrations, one can see their effect on the E_H. For example, the last term can be used to illustrate the magnitude of the change in the E_H when the reaction is considered at pH 7 ($[H^+]=10^{-7}$) instead of at standard conditions ($[H^+]=1$):

$$\frac{RT}{8F} \ln [10^{-7}]^{10} = -\frac{70RT}{8F} \ln 10 = -0.52\,\text{V} \quad (T = 298\,\text{K}) \qquad (3.19)$$

This correction is large enough to change the sign of the E_H for the half reaction. Corrections due to species concentrations being different from the standard concentration of 1 M can be considered in a similar way. Figure 3.4 illustrates the values of the E_H at pH 7 (and $T = 4°C$, rather than $25°C$) at standard concentrations (1 M), and for the more environmentally relevant concentrations of the reactants and products. Note that, for redox pairs arranged this way in Figure 3.4, a redox reaction is thermodynamically favorable when the pair containing the oxidant (on the left side of the pair: O_2, SO_4^{2-}, etc.) is located higher on the E_H diagram than the pair containing the reductant (on the right side of the pair: NH_4^+, CH_4, etc.). Thus, sulfate can be reduced by methane but not by ammonium.

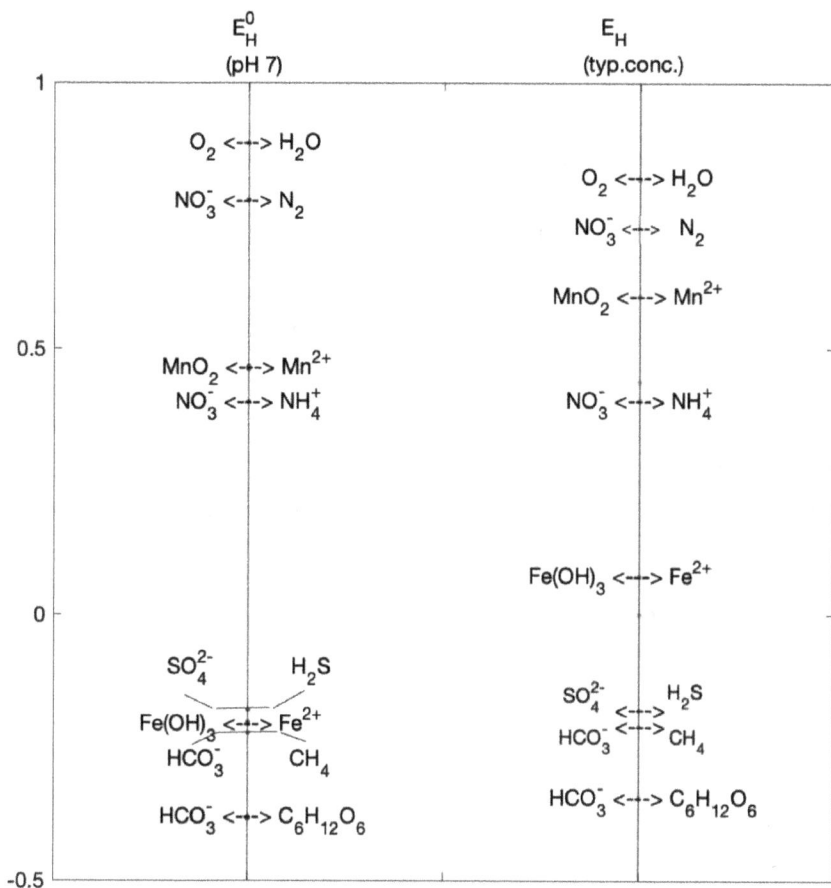

Fig. 3.4. E_H values for common diagenetic half-reactions, corrected to pH 7 and $T = 4°C$. Redox reactions are favorable if the E_H for the oxidant (the species at the left side of each pair) is greater than the E_H for the reductant (on the right side of the pair). (Left) For standard concentrations of 1 M (except for $[H^+]$). (Right) For environmentally relevant concentrations: $[HCO_3^-] = 2$ mM; all other dissolved species are at 10 μM.

Test your understanding: Calculate the change in the E_H for sulfate (SO_4^{2-}) reduction to sulfide (H_2S) that would be induced by a decrease in pH by 1 unit. What would be the corresponding change in the ΔG for a reaction in which sulfate is used to oxidize acetate $(C_2H_3O_2^-)$? (See Table C.6 for half reactions.)

3.5   Mineral Precipitation and Dissolution

The energetics of mineral precipitation follows the same thermodynamic framework as that for the chemical reactions discussed above, but the process is worth discussing in more detail. Creation of mineral crystals involves forming an initial nucleus, followed by its subsequent growth by attachment of atoms sourced from the surrounding solution.

For very small crystals, the energy of attaching an atom or molecule from solution to the growing crystal depends on the properties of the mineral surface. As work needs to be done against surface tension, forming the initial nucleus requires more work than adding subsequent atoms to it (Fig. 3.5). For this reason crystals grow easier on already existing particles, which could serve as crystallization centers. This phenomenon is not unlike the bubbles in a tea kettle forming first on the kettle walls well before boiling can spread to the entire water volume. Microbial cells are also large enough to facilitate precipitation on their surfaces, a phenomenon that is exploited to their advantage by a range of microbes whose metabolisms require them to deal with solid substrates.

Once the initial barrier to nucleation is overcome, the positive energy gain (negative ΔG) means that the attachment of atoms to the crystal surface is easier (i.e. more frequent) than detachment. Crystal growth then occurs as long as there is energy gain, which depends on the concentrations of reactants analogously to Eq. (3.4):

$$\Delta G = \Delta G^0 + RT \ln Q = RT \ln \frac{Q}{K_{eq}} \qquad (3.20)$$

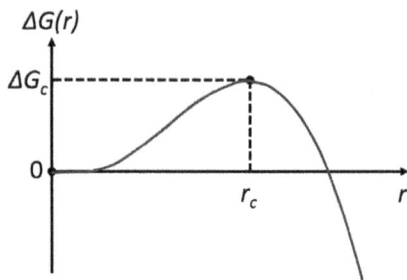

Fig. 3.5.   Dependence of the Gibbs free energy of crystallization on crystal size. Below a certain critical size r_c, growth of crystals is unfavorable (increases free energy).

Here, Q (sometimes also denoted Ω) is the activity product of ions in solution, and K_{eq} (called the *solubility product constant*; Table C.3) is the activity product at equilibrium, when the concentrations are such that the energy for crystal growth ΔG is zero:

$$K_{eq} = e^{-\frac{\Delta G^0}{RT}} \qquad (3.21)$$

The ratio of the activity products in Eq. (3.20) (called the *saturation ratio*) then defines the thermodynamic condition for crystal growth: the growth is favorable ($\Delta G < 0$) when the ratio is greater than 1 ($Q > K_{eq}$), and dissolution is favored if it is less than 1. For example, precipitation of solid calcium carbonate is described by the reaction

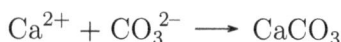

$$Ca^{2+} + CO_3^{2-} \longrightarrow CaCO_3$$

The activities of solid species, by convention, are equal to 1. So, for growth, the concentrations of the calcium and carbonate ions must satisfy the equation:

$$\frac{Q}{K_{eq}} = \frac{[Ca^{2+}][CO_3^{2-}]}{K_{eq}} > 1$$

Solubility constants K_{eq} for common minerals are found experimentally and tabulated (Table C.3). The logarithm of the ratio, which shows how much the actual concentrations exceed the equilibrium values,

$$SI = \log Q/K_{eq} \qquad (3.22)$$

is termed the *saturation index*. At equilibrium, the ratio Q/K_{eq} equals 1, and its logarithm is 0. Dissolution occurs when the saturation index falls below zero, whereas growth occurs for the SI above zero. Note, however, that the saturation index reflects only the thermodynamic condition, i.e., whether mineral formation is energetically favorable. Whether precipitation and growth actually occur depends also on the speed at which the reaction can proceed, i.e., its kinetics (see below).

Example: Calcium carbonate precipitation. What conditions favor the formation of calcium carbonate? Let us consider a range of typical concentrations and pH values. Calcium carbonates include several minerals, the most common being calcite and aragonite, which share the same chemical formula $CaCO_3$ (Appendix B). We will assume calcite for this example. The solubility constant is listed in Table C.3: $K_{eq} = 10^{-8.48}$ (at T=25°C). Let us consider two typical Ca^{2+} concentrations: 10 mM (close to that in seawater) and 1 mM (close to that in Lake Michigan). The concentration of CO_3^{2-} at which supersaturation is achieved is

$$[CO_3^{2-}] > \frac{10^{-8.48}}{[Ca^{2+}]}$$

where the calcium concentration should be in M. This gives the CO_3^{2-} concentrations of, respectively, $10^{-6.48}$ M $= 0.3\,\mu$M and $10^{-5.48}$ M $= 3.3\,\mu$M, for the marine and freshwater Ca^{2+} concentrations, at standard pH 7. The total amount of dissolved carbon dioxide (DIC) can be calculated from Eq. (C.3): at pH 7, it gives, respectively, 0.785 mM and 7.85 mM. For typical DIC concentrations of a few mM, these numbers would suggest that seawater should be usually supersaturated with respect to calcite, while lake water should be undersaturated. The results are strongly pH dependent, however. Figure 3.6 shows the result from Eq. (C.3) over a range of pH values. Lake water may become supersaturated at higher pH: the surface water of Lake Michigan, for example, has the pH of about 8.5, which

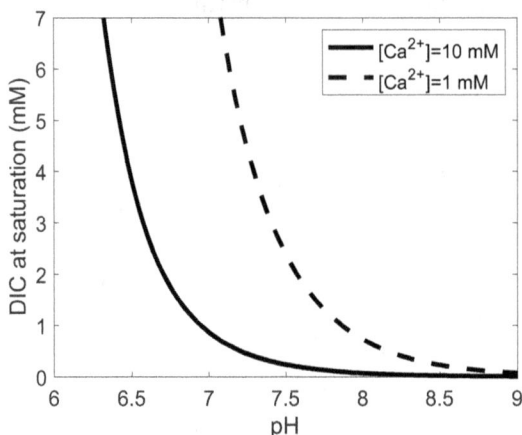

Fig. 3.6. Concentration of total dissolved CO_2 (termed "dissolved inorganic carbon", DIC) at saturation equilibrium with respect to calcite. (At T $= 25$°C.)

is squarely in the supersaturation range. At low pH (which may be achieved in sediment, for example) even seawater may become undersaturated with respect to calcite, which may lead to calcite dissolution. These conditions for calcite precipitation and dissolution are very important for aquatic organisms that use calcium to build their shells or exoskeletons, as well as for the preservation of the remains of such organisms in the sediments.

3.6 Kinetics of Chemical Reactions

The thermodynamic favorability of a reaction, while determining its direction, says very little about the rate at which the reaction is to proceed. Oxidation of cellulose by O_2, for instance, is a highly favorable reaction even at room temperature, but wood does not spontaneously catch fire. Microbial consortia in soil, however, are capable of slowly breaking down wood, extracting energy for their metabolisms in the process. While energetic favorability is dictated by rigorous and well-established laws of thermodynamics, the rates (kinetics) of reactions are influenced by a multitude of factors, many of which are poorly known and often inferred only empirically. In diagenetic settings, some reactions, including microbially-catalyzed reactions close to thermodynamic equilibrium, proceed slowly, taking years or even millennia. Others, like inorganic acid-base equilibrations, proceed exceptionally quickly and achieve equilibria within a fraction of a second.

3.6.1 Forward and reverse reactions

The thermodynamic favorability given by ΔG describes the direction of the net reaction. But individual molecules may undergo transformations in either forward of reverse directions, albeit with different probabilities. Energetically favorable transitions occur more frequently than their opposites. The rate of the net reaction (r) represents the difference between the rates of forward (r_+) and reverse (r_-) reactions:

$$r = r_+ - r_-$$

(3.23)

Reactions that are thermodynamically favorable may not happen spontaneously if the process requires overcoming an *activation*

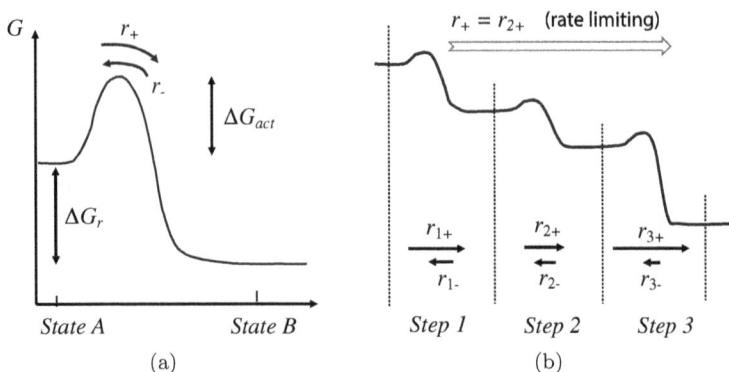

Fig. 3.7. (a) The reaction rate is the net result of the forward and backward reactions. (b) Multi-step reactions. Individual steps may proceed at different rates.

barrier (Fig. 3.7). Providing additional energy to overcome that barrier can make the reaction proceed spontaneously, with net energy gain. Igniting wood to initiate its highly energetic oxidation is an example. Many microbial enzymes that catalyze reactions in the environment function to move electrons across energetic barriers, enabling reactions that otherwise would not spontaneously occur.

When a reaction proceeds through a series of steps, individual steps by themselves may, in principle, proceed at different rates (Fig. 3.7). The overall reaction, however, may only proceed as fast as its slowest step. The rate of the overall reaction (e.g., the number of molecules transformed per unit volume per unit time) then becomes defined by the rate of its *rate-limiting* step.

The kinetics of reactions that occur far from thermodynamic equilibrium are often characterized (experimentally, as well as mathematically) using descriptions that reflect only the forward reaction, as the reverse reaction is negligible. In diagenesis, however, much slower kinetics near equilibrium routinely become important. Nonequilibrium thermodynamics predicts that the ratio of the forward and reverse rates depends on the Gibbs free energy of the reaction as

$$\frac{r_+}{r_-} = e^{-\frac{\Delta G}{\chi RT}} \tag{3.24}$$

Here, χ is a stoichiometric parameter, which describes the number of times that a rate-determining step occurs per turnover of the reaction

[26]. The reaction proceeds forward when ΔG is negative, and in reverse when it is positive. Rewriting ΔG in Eq. 3.24 via the activity product (Eq. 3.20), the net rate $r = r_+ - r_-$ can be written as

$$r = r_+ \left[1 - \left(\frac{Q}{K_{eq}} \right)^{1/\chi} \right] \qquad (3.25)$$

This indicates the functional form for the dependence of the net rate on the degree of thermodynamic disequilibrium, and how it may differ from the rate of the forward reaction alone [5]. The forward rate r_+ in Eq. 3.25 is determined by kinetic factors.

3.6.2 Rate laws

Except for the simplest reactions, for which theoretical calculations may be possible, kinetic expressions for the rates of chemical reactions are obtained empirically as functions of species concentrations. Such functional expressions are known as *rate laws*. (Despite the name, these "laws" do not carry the same authority as the laws of gravity or energy conservation.) Theoretically, for a reaction that takes place in bulk solution (a homogeneous reaction), the reaction rate could be expected to be proportional to the probability of encountering the reactant molecule in solution, i.e., to the concentration of the reactant. Then the probability of two molecules combining in a reaction would be proportional to the product of their concentrations. In practice, reaction rates can be regulated by multiple factors, so their rate laws are inferred phenomenologically. Simple formulations include:

$$\text{Zero order:} \quad \frac{dC}{dt} = -k_0 \qquad (3.26)$$

$$\text{First order:} \quad \frac{dC}{dt} = -k_1 C \qquad (3.27)$$

$$\text{Second order:} \quad \frac{dC_A}{dt} = -k_2 C_A C_B \qquad (3.28)$$

Here, k_0, k_1, k_2 are reaction rate constants, and C are concentrations. As the rate law reflects essentially the slowest, rate-limiting step in the reaction, the order of the rate law does not necessarily need to reflect the actual number of reactants.

Reactions catalyzed by microbes often follow more complicated kinetics. Catalytic reactions, including those facilitated by microbial enzymes, are commonly described by the *Michaelis–Menten* type of rate law

$$R = \frac{dC}{dt} = V_{\max}\frac{C}{C + K_m} \tag{3.29}$$

At high concentrations C, the rate R is not limited by the concentration of the reactant but is instead limited by the kinetics of the microbial metabolism, so the rate reaches a maximum $R = V_{\max}$ when $C \gg K_m$. The half-saturation constant K_m corresponds to the concentration when the rate is half of its maximum value. At low concentrations $C \ll K_m$, the reaction is limited by the reactant and the rate becomes linearly proportional to the concentration: $R \approx V_{\max}C/K_m$. Importantly, at very low concentrations, microbially-catalyzed reactions may cease altogether, if the energy that a microbes extracts from the reaction is insufficient to offset the metabolic costs for conducting such a reaction (synthesizing enzymes, etc.). Reaction rate laws for such reactions then need to account for such thresholds, often by multiplying the rate expression by a *thermodynamic factor*, which becomes zero below the concentration threshold. More detailed descriptions of the kinetic controls and their mathematical descriptions are given in Chapters 4 and 11.

The temperature dependence of a reaction rate often follows the Arrhenius exponential dependence:

$$k = Ae^{-\frac{\Delta E}{RT}} \tag{3.30}$$

where T is the absolute temperature (in degrees Kelvin) and ΔE is the activation energy of the reaction. In practice, ΔE and A are often treated as empirical coefficients.

Test your understanding: Figure 3.8 shows an experiment for determining the kinetics of a reaction. What is the best approximation for the rate law? What is the rate constant?

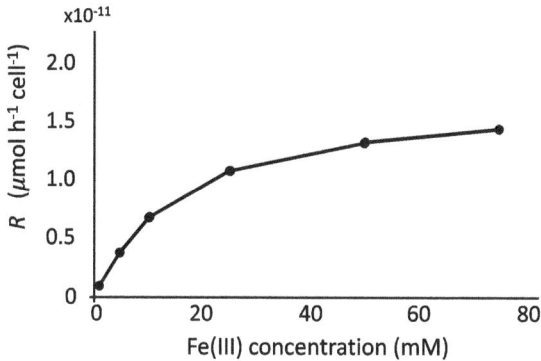

Fig. 3.8. Reaction kinetics in an incubation experiment (from [27]). The rate of microbial reduction of lepidocrocite (an iron oxyhydroxide mineral) was measured as a function of the mineral concentration. The rate here is normalized by the number of microbial cells of the organism that performs the reduction.

3.6.3 Kinetics of mineral precipitation and dissolution

Reactions of mineral precipitation or dissolution are, by definition, heterogeneous (involving more than one phase) and involve the several steps with potentially different rates: diffusion of reactants to mineral surface, diffusion of products away from the surface, and the chemical reaction (creation or breaking of chemical bonds) at the surface. The kinetics of attachment and detachment of molecules at the surface may also be important, though they are typically rapid and unlikely to limit the rate of the overall reaction.

When the reaction at the mineral surface is rapid, the rate of the overall reaction may be limited by the speed at which the reactants and products are transported to or from the surface by diffusion. The reaction is then said to be *diffusion-controlled*. If diffusion is rapid but the reaction at the surface is slow, the overall reaction is said to be *surface-controlled*. As diffusion is rapid over the short distances comparable to the sizes of mineral grains (Table 2.2), many diagenetic reactions of mineral precipitation and dissolution are surface-controlled [5].

A common theoretical approach [5] to describing surface-controlled reactions is through a so-called transition state theory,

surface-controlled

slow   fast

diffusion-controlled

fast   slow

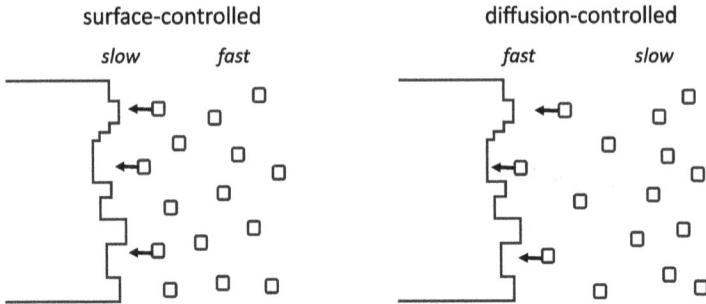

Fig. 3.9. Diagram illustrating the distribution of reactants in a surface-controlled vs. diffusion-controlled heterogeneous reaction.

which posits the creation of an "activated complex", which can overcome the surface potential barrier. The dissolution rate then depends on the rate at which the complex decays and is independent of the saturation in the bulk solution. The opposite reaction, precipitation, depends on the supersaturation. The net rate of the overall reaction is set by the balance between precipitation and dissolution. The rates of the two processes become equal at equilibrium. Such kinetic equilibrium, in principle, need not occur at the same ambient conditions as thermodynamic equilibrium. While no growth can occur if the bulk solution is not oversaturated, thermodynamic oversaturation, by itself, may not be sufficient for growth. The degree of oversaturation thus determines both the thermodynamic favorability and the rate of mineral growth. The degree of undersaturation, on the other hand, while defining a thermodynamic condition for dissolution, may not exert the same degree of control over the speed at which the mineral dissolves. Specific mathematical expressions that may be used to describe the kinetics of mineral growth or dissolution are listed in Chapter 11.

3.6.4   Reaction-controlled vs. transport-controlled kinetics

In vertically stratified sediments, the macroscopic outcomes of chemical reactions depend not only on the reaction's intrinsic kinetics but also on the rates of physical processes that transport the reaction's reactants and products. Reactants are brought into the reaction zone, and reaction products are similarly removed from the reaction zone, maintaining thermodynamic disequilibrium. When the chemical kinetics is fast, the overall rate of consumption (or production) of chemical species may become limited by the rates of their physical transport to (or from) the reaction zone. The reaction rate, in that case, becomes *transport-controlled*. Conversely, for slow reactions, while physical transport may be providing sufficient amounts of molecules, the overall rate can be only as fast as the chemical reaction. The process is then said to be *reaction-controlled*.

For example, in marine sediments, sulfate (SO_4^{2-}) can react with methane (CH_4) in a process known as the anaerobic oxidation of methane (AOM). As sulfate is drawn from the overlying water while methane diffuses upward from deep sediment, the reaction takes place within a relatively narrow zone where the two species meet. Often

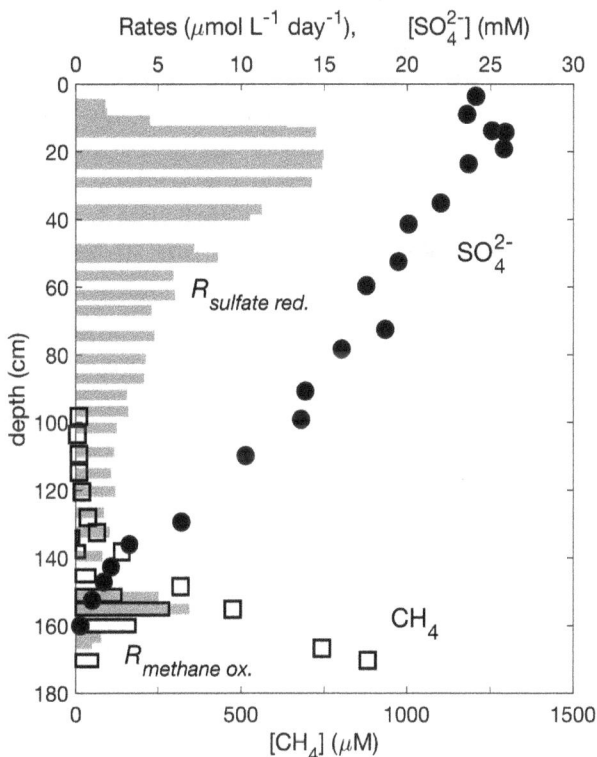

Fig. 3.10. Profiles of sulfate and methane in marine sediment from the Kattegat, and the associated rates of sulfate reduction and methane oxidation. Redrawn from [28] and [29]. The AOM reaction takes place in a narrow zone about 150 cm below the sediment surface.

this happens several meters below the sediment surface (Fig. 3.10). If the (microbially-catalyzed) reaction between sulfate and methane can happen faster than the diffusion of sulfate or methane over such distances (Table 2.2), then the overall rate of AOM becomes limited by diffusion (of either sulfate or methane, or both). To quantitatively compare the relative speeds of these processes, we can compare their characteristic time scales. The diffusive transport of sulfate takes several decades (Table 2.2). The reaction time scale can be estimated from the reaction kinetics. For AOM, the reaction takes place in bulk porewater and relies on the supplies of both sulfate and methane; accordingly, it is often considered second-order:

$$R = k[SO_4^{2-}][CH_4]$$

For the reaction rate constant k, some studies considered values as high as 8000 $M^{-1} y^{-1}$ [30]. For this value, even for the concentrations in the reaction zone as low as 0.1 mM, the time scale of the reaction is very short: $0.1 \times 10^{-3}/8000 = 13 \times 10^{-9}y = 0.4\,sec$ (assuming that the rate is not limited thermodynamically or by microbial energetics, which is often the case [30] and can be verified separately). The rate at which methane becomes anaerobically oxidized in the sediment then becomes limited by diffusion. (In turn, the vertical flux of the upward-diffusing methane is set by the rate at which methane is produced in the deep sediment from organic matter; so the total rate of AOM can be nearly equivalent to the total, vertically integrated, rate of methanogenesis.)

Test your understanding: Organic particles a few cm below the sediment-water interface are characterized by a typical first-order reactivity constant of about $k = 0.01\,y^{-1}$, meaning that a substantial fraction of particulate organic carbon is broken down over the time scale $\tau = 1/k = 100$ years. If organic carbon is being oxidized by oxygen supplied into the sediment from the overlying water by molecular diffusion, is the rate of the mineralization reaction controlled by the kinetics of the reaction or by transport?

3.7   Shapes of Vertical Concentration Profiles

Vertical profiles – variations in chemical concentrations with depth in the sediment – are probably the most common and basic

characterizations of sediment geochemistry. Chemical reactions act
to increase or decrease the concentrations of reactive species within
sediment layers. Vertical variations from one layer to another, however, reflect not only the reactions but also physical transport. It is
thus helpful to recognize how the interplay of these processes becomes
reflected in the profile shapes.

When a sediment layer serves as a net source of a chemical
species (i.e., the chemical is being produced there), its concentration
increases relative to the neighboring layers. For example, ammonium
(NH_4^+) is being produced as organic matter decays within the sediment, so its concentration below the sediment surface is higher than
in the water column (Fig. 3.12). Ammonium continues to be produced as organic matter is buried into the deeper sediment, so its
concentration continually increases with depth. The rate of production, however, slows down with depth, as the pool of available organic
matter becomes progressively exhausted with time after deposition.
The increase in ammonium concentration from one layer to the next
therefore becomes smaller, hence the slope of the ammonium profile decreases. Similar shapes would be associated with other species
being generated within the sediment, such as methane (CH_4) or dissolved ferrous iron (Fe^{2+}) (Fig. 3.12). Locations of the steepest slopes
are associated with the highest rates of production. A convex profile indicates a net production of the chemical within the layer (Fig.
3.11). A concave profile indicates consumption. (See also Fig. 11.2.)
Importantly, a peak in a chemical profile (indicating a source of production) cannot be maintained indefinitely against diffusion unless it
is bounded on both sides by layers that serve as sinks for the chemical
(Fig. 3.11). In the absence of sinks, diffusion would lead to a flatter
profile, such as at the lower end of the Fe^{2+} profile in Fig. 3.12. Likewise, a dip in the concentration profile that remains stable over time
would indicate a sink for the chemical that is bounded on both sides
by sources (Fig. 3.11). A more rigorous and more complete mathematical description is given in Section 11.2.2 and summarized in
Fig. 11.2.

Test your understanding: What does the shape of the sulfate
(SO_4^{2-}) profile in Fig. 3.12 indicate about the net production or consumption of this species in this sediment?

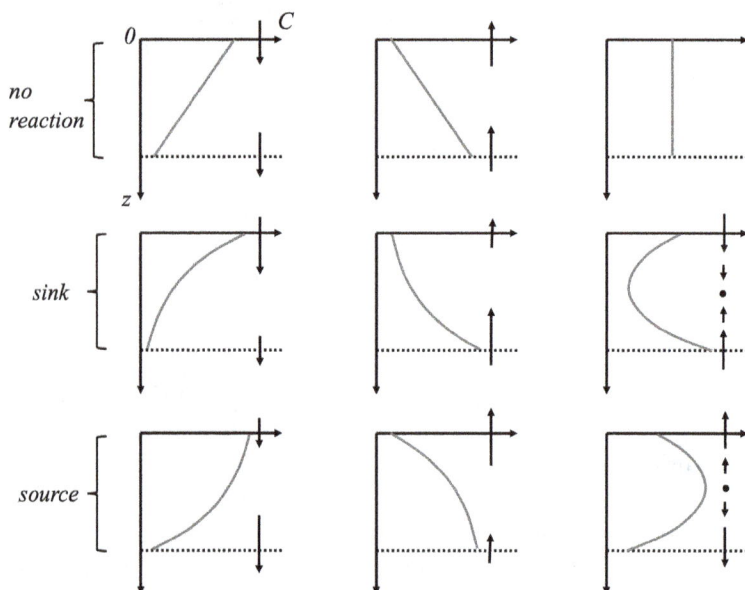

Fig. 3.11. Shapes of vertical profiles generated by reactive sources and sinks within the sediment. Arrows indicate the direction and magnitude of vertical transport fluxes, driven by chemical gradients. See also Fig. 11.2 for more explanation.

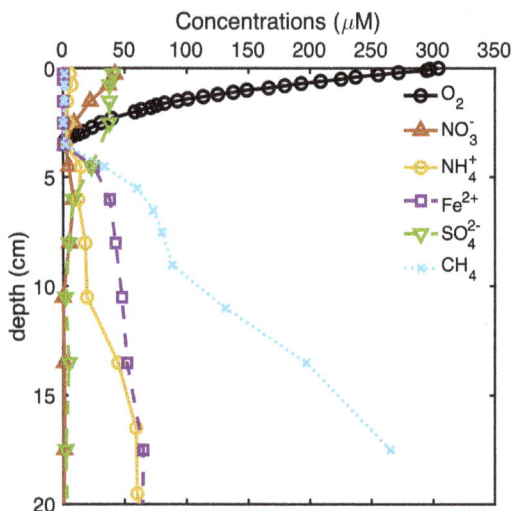

Fig. 3.12. Profiles of multiple chemical species in the porewater of sediment from a large oligotrophic lake (Lake Superior [11, 31, 32]). The shapes of the concentration profiles reflect the sources and sinks of the chemical species within the sediment.

3.8   Exercises

1. A nutrient is evenly distributed through a 20 m water column at a concentration of 3 $\mu mol\,L^{-1}$. What is the integrated amount of nutrient per square meter?

2. Organic carbon is produced by photosynthesis from atmospheric CO_2 in the upper 1 m of the lake at a rate of 5 $\mu molC\,L^{-1}d^{-1}$. If this production continues for a month and the produced organic matter then all settles to the lake bottom, what would be the depositional flux of carbon? (Obviously, this exercise on unit conversion is not a realistic ecological example, as it ignores mineralization and respiration.)

3. To what pH does one need to acidify samples containing dissolved Fe^{2+} to prevent the oxidation of iron to FeOOH?

4. The concentration of dissolved Fe^{2+} in the sediment porewater is 10 μM, at neutral pH. What concentration of the hydrogen sulfide ion HS^- can coexist in this porewater without causing precipitation of iron monosulfide (FeS)? What is the corresponding concentration of total dissolved hydrogen sulfide (including H_2S)?

5. At what pH would the precipitation of calcium carbonate be expected in a lake where the concentrations of calcium and DIC are, respectively, 0.5 mM and 3 mM? (In lakes, events of such spontaneous precipitation caused by increases in pH during periods of intense photosynthesis are called "whiting events", as they turn the water milky white.)

6. What should be the concentration of methane in the sediment at a depth of 500 m to enable growth of methane bubbles? How about the DIC concentration for CO_2 bubbles?

7. Hydrogen sulfide diffuses upward from anoxic sediment and is being oxidized by oxygen in a thin sediment layer where the concentration of oxygen is approximately 1 μM and the concentration of hydrogen sulfide is also approximately 1 μM. The second-order reaction rate constant is $k = 1 \times 10^7 M^{-1}y^{-1}$. What is the reaction rate? Would the rate of hydrogen sulfide oxidation in this sediment be expected to be reaction-controlled or diffusion-controlled?

8. What does the shape of the nitrate profile in Fig. 3.12 indicate about nitrate production or consumption? Does the water column serve as a sink or a source of nitrate for the sediment? Nitrate can be produced within the sediment when ammonium is oxidized by

oxygen within the oxidized upper sediment layer. Assuming that the oxidation reaction is fast so that the overall rate of ammonium oxidation is limited by the rate of ammonium diffusion from deeper sediment, estimate the rate of ammonium oxidation based on the upward flux of ammonium.

Chapter 4

Microbial Processes

This chapter discusses the contributions of microbes to geochemical reactions and elemental cycling. Microbes catalyze most of the diagenetically relevant reactions, in the process extracting energy for their metabolisms. The geomicrobiological literature is vast and rapidly evolving, and it would be impossible to cover all of relevant topics in this brief review. Here, we focus only on those aspects that are directly linked to the material in other chapters of this book. We review the terminology that is commonly used describe to microbial populations and their metabolisms, then summarize the differences between respiratory and fermentative catabolisms and how they relate to the redox reactions discussed in the previous chapter. Finally, we address several additional aspects of microbial ecology and give a very brief introduction to the array of molecular genetics tools that have emerged in the last several decades.

4.1 The Basics

The terms "microorganism" or "microbe" commonly refer to prokaryotic single-cell organisms, which fall into two large groups (domains): *Bacteria* and *Archaea*. Together, they represent by far the greatest share of organisms on Earth, dwarfing eucaryotic organisms in both genetic diversity and total biomass. Most of the microorganisms have cell sizes on the order of 1–10 μm, but giants up to 1 mm in size can also be found in both marine and freshwater sediments [33,34]. Classification schemes nowadays are commonly based on genomic sequencing information. For our geochemical purposes, however, distinctions based on the type of environment (e.g., presence or absence of oxygen, aerobes vs anaerobes) or the geochemical function (e.g., Fe reducers) are often more important.

4.2 Microbial Metabolisms

Microorganisms are biological machines that run on energy they draw from the environment. Four billion years of evolution have equipped them with a multitude of ways to extract that energy. In sediments where sunlight is unavailable, the energy comes from chemical reactions. As microbes adapt to their environment, it is a reasonable general principle that, for reactions that release chemical energy, one may expect to find corresponding microbial populations that use it, with their specific sets of enzymes and other intracellular and extracellular machinery. Their effectiveness is governed by thermodynamics, chemical kinetics, and competition with other microbial populations and abiotic reactions that use the same substrates. Biological constraints, such as the physiological capabilities of the microbes or their tolerance to toxins, also have important effects. Some general principles, nevertheless, can be gleaned from the overall thermodynamic setting, without going into too much biological detail. In particular, higher densities of available chemical energy – for example in the form of reactive organic matter support higher densities of microorganisms and higher rates of their chemistry-based metabolisms. The very top millimeters of sediments may contain several years worth of organic sedimentation. Accordingly, the population densities of microbial cells there can be several orders of magnitude greater than those in the overlying water column. These densities decline with depth into the sediment, as energy becomes less available (Fig. 4.1).

4.2.1 Basic terminology

Fundamentally, most microbial metabolisms involve catalyzing an exergonic chemical reaction, converting some of its energy into the chemical energy of the adenosine triphosphate molecule (ATP, which functions as the cell's "currency" of energy transfer), and then using the ATP to synthesize organic molecules that make up the cell. As a

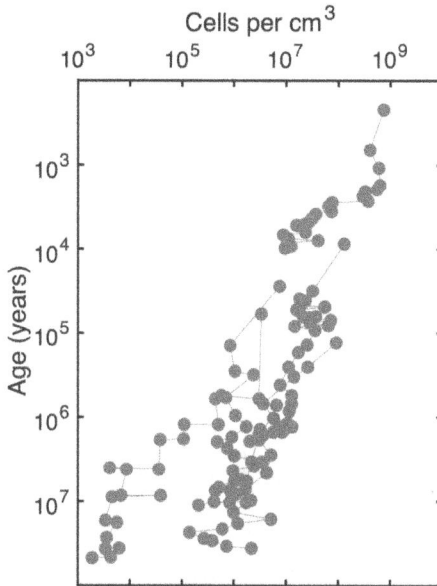

Fig. 4.1. Concentrations of microbial cells as a function of depth within the sediment (age) in a range of marine sediments (redrawn from [35]). The surface of the sediment contains around 10^9 cells per cm^3. The numbers decrease with depth, as the amount of labile organic matter decreases, but active microbial populations persist well into the deep biosphere.

source of carbon for these molecules, microbes can use either ambient CO_2 or organic compounds sourced from their environment. Some energy has to be expended to maintain the cell's internal environment (homeostasis), and some is inevitably dissipated into the environment as heat. Obtaining energy from the environment is termed *catabolism*. Synthesizing new organic molecules for biomass is termed *anabolism*. Together, these two processes constitute *metabolism* (Fig. 4.2).

Metabolisms thus require several generic components: a source of energy and a source of carbon (in addition to other, less abundant, elements such as N and P) for building biomass. As energy is generated by transferring electrons between atoms, additional chemical compounds may also be required as oxidants or reductants. These components form the basis of a commonly used classification of microorganisms by their metabolic function. Metabolism names are constructed by listing the corresponding sources of energy, electrons, and carbon – in that order, as illustrated in Fig. 4.3 and Table 4.1.

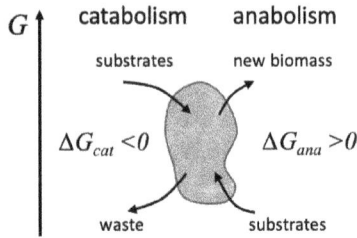

Fig. 4.2. Metabolism is made up of catabolism and anabolism. Catabolism produces energy for the cell; anabolism uses the energy to build new biomass (redrawn from [36]).

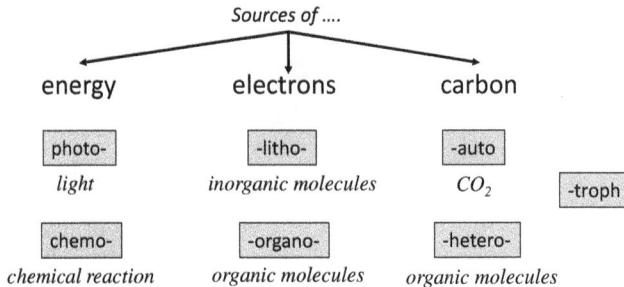

Fig. 4.3. Naming of microbial metabolisms.

For example, a sulfate reducer may obtain its energy from a reaction that reduces sulfate (SO_4^{2-}) by using lactate ($C_3H_4O_3^-$) as an electron donor. The oxidation reaction produces acetate ($C_2H_3O_2^-$). The organism then uses this acetate as a building block for biosynthesis. Such an organism would be called a chemo-organo-heterotroph. An organism that obtains its energy by oxidizing hydrogen sulfide (H_2S) with oxygen (O_2) and uses that energy to synthesize its biomass by pulling CO_2 from the environment is a chemolithoautotroph. Shorter names that use only one or two of the three categories can be also constructed: chemotroph, lithotroph, heterotroph, chemolithotroph, etc.

When a chemical produced in a catabolic reaction is further used in biosynthesis, the corresponding metabolic pathway is termed *assimilatory*. For example, in an assimilatory nitrate reduction, nitrate (NO_3^-) is reduced within the cell to ammonium (NH_4^+), which is then used in biosynthesis as a source of nitrogen. If the reactant is instead rejected from the cell to the environment, the pathway is

Table 4.1. Examples of microbial metabolisms (after [37]).

Organism	Energy	e^-	Carbon	Metabolism
Algae, cyanobacteria	light	H_2O	CO_2	Photolithoautotroph
Nitrifying bacteria	NH_4^+ ox.	NH_4^+	CO_2	Chemolithoautotroph
Methanogenic archaea (hydrogenotrophic)	H_2 ox.	H_2	CO_2	Chemolithoautotroph
Denitrifying bacteria	C_{org} ox.	C_{org}	C_{org}	Chemoorganoheterotroph
Mn- and Fe-reducing bacteria	C_{org} ox.	C_{org}	C_{org}	Chemoorganoheterotroph
Sulfate reducing bacteria	C_{org} ox.	C_{org}	C_{org}	Chemoorganoheterotroph
Methanogenic archaea (acetoclastic)	Ac ox.	C_{org}	C_{org}	Chemoorganoheterotroph
Fermenting bacteria	C_{org} ox.	C_{org}	C_{org}	Chemoorganoheterotroph
Methane oxidizing bacteria	C_{org} ox.	CH_4	C_{org} (or CO_2)	Chemoorganoheterotroph

termed *dissimilatory*. For example, the dissimilatory nitrate reduction to ammonium (DNRA) is a chemotrophic pathway that can produce biologically available nitrogen (as ammonium) within the sediment.

Organisms that use oxygen in their catabolisms are *aerobes*, whereas those inhabiting the anoxic deeper sediment are *anaerobes*. Many microorganisms are capable of multiple catabolisms. When an organism (such as a denitrifying bacterium) does not require oxygen but would use it when given a choice, it would be termed a *facultative* aerobe. Those with a rigid requirement for oxygen would be *obligate* aerobes.

Test your understanding: Where in the sediment would you most likely find favorable conditions for chemolithoautotrophs? For photolithotrophs?

4.2.2 Respiration and fermentation

Cells derive energy for their metabolisms through chemical reactions, i.e., by transferring electrons between atoms. In organotrophic metabolisms, electrons are transferred from electron donors such as acetate to electron acceptors such as oxygen (or others, see

Chapter 5). From a biological perspective, this process is *respiration*. (More generally, respiration is oxidation of organic carbon with external electron acceptors [6].)

An alternative way to procure energy, which does not involve an external electron acceptor, is *fermentation*. In fermentation, electrons are transferred between parts of the same organic molecule, which serves as both the electron donor and the electron acceptor. In the process, the organic molecule is cleaved into smaller moieties. For example, fermentation of acetate (CH_3COOH) transfers electrons between the two carbon atoms, generating both reduced (CH_4) and oxidized (CO_2) carbon compounds:

$$CH_3COOH \longrightarrow CH_4 + CO_2 \qquad (4.1)$$

The large variety of organic molecules offers numerous possibilities for fermentation reactions. For example, fermentation of glucose can produce acetate

$$C_6H_{12}O_6 + 2\,H_2O \longrightarrow 2\,C_2H_3O_2^- + 2\,CO_2 + 4\,H_2 + 2\,H^+$$
$$\Delta G^0 = -50\,kJ/mol$$

lactate

$$C_6H_{12}O_6 \longrightarrow 2\,C_3H_4O_3^- + 2\,H^+ \quad \Delta G^0 = +55\,kJ/mol$$

or ethanol

$$C_6H_{12}O_6 \longrightarrow 2\,C_2H_6O + 2\,CO_2 \quad \Delta G^0 = -206\,kJ/mol$$

(the last reaction, in addition to powering microbes, also powers some microbiologists).

The energy derived from respiration or fermentation is stored within the cell as ATP (adenosine triphosphate), which then can be used for cell maintenance and reproduction. In respiration, electrons are transferred from the electron donor when it is bound in a complex on the cell membrane. They are then transferred within the membrane through the *electron transport chain* of enzyme complexes and cofactors to the electron acceptor (Fig. 4.4). In the process, hydrogen ions are driven from the cell's cytoplasm into the environment (*proton translocation*), and re-enter the cell through a

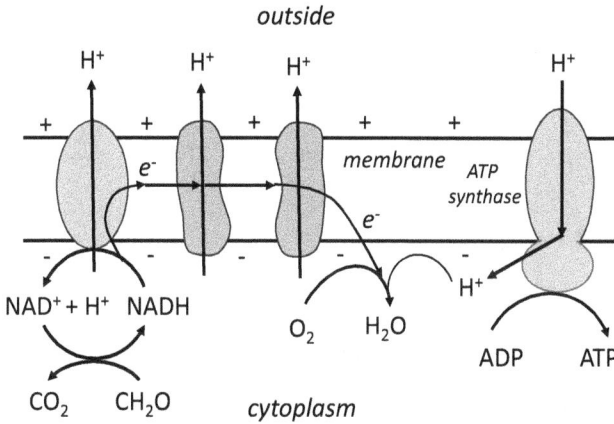

Fig. 4.4. The electron transport chain across the cell membrane. Electrons are produced by the oxidation of a reduced electron donor, e.g., an organic compound (CH_2O). The electron transport chain is a series of redox couples that transport the electron, while discharging protons to the outside of the membrane. The electron is consumed by the reduction of an electron acceptor, such as O_2. ATP is generated when protons are passed back inside through the ATP synthase.

special membrane-bound enzyme *ATP synthase*, in which ATP is produced from adenosine diphosphate (ADP) and orthophosphate ions [5]. In fermentation, the electron donor and the electron acceptor may be the same molecule, so the electron transfer may occur at the same place on the cell membrane.

4.2.3 Fermentation and respiration during organic matter breakdown

In microbial ecosystems, fermenting organisms perform the important function of breaking down complex organic molecules into smaller compounds, which can be then metabolized by other microbial groups. Respiring organisms, especially anaerobes such as iron reducers or sulfate reducers, typically can access only low-molecular-weight (LMW) compounds, such as acetate. Their metabolisms in sediments therefore rely on the presence of fermenters, which can both hydrolyze particulate organic material and break down the resulting high-molecular-weight (HMW) compounds into simpler ones (Fig. 4.5). Aerobic microbes are typically able to perform such steps for themselves, which gives them an additional advantage over

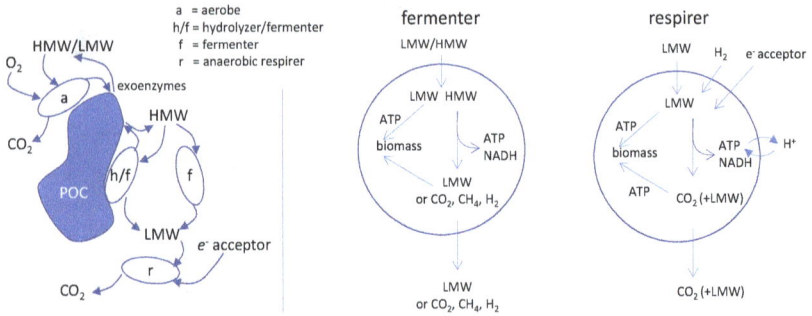

Fig. 4.5. Fermentation and respiratory metabolisms in the mineralization of organic matter (after [6]).
Note: HMW = high-molecular-weight organic molecules (components of dissolved organic carbon, DOC); LMW = low-molecular-weight organic molecules (also part of DOC); POC = particulate organic carbon.

anaerobes. Most aerobic heterotrophs are capable of breaking down organic substrates completely into CO_2, water, and nutrients [6].

The metabolic activities of aerobes strongly affect microbial ecosystems. A side effect of using oxygen is the formation of highly reactive oxygen radicals, such as superoxide anion, hydrogen peroxide, and hydroxyl radicals. They can break chemical bonds in compounds that are normally refractory, such as lignin, making the decomposition of organic material more complete. They can also harm cellular components of other cells, which makes aerobes toxic to many anaerobic organisms. This leads to a fairly strict spatial separation between the sediment zones of aerobic and anaerobic matabolisms.

4.2.4   Hydrogenotrophy

Many fermentation reactions produce molecular hydrogen (H_2), which is an excellent electron donor that can support microbial metabolisms. Hydrogen can be oxidized instead of organic electron donors in the catabolisms of manganese reducers, iron reducers, sulfate reducers, methanogens, and other diagenetically important microbial groups (Table D.1). Microbial populations in sediments compete for this valuable resource. Depleting its concentration below the threshold of thermodynamic favorability for the less efficient competitors is a way to shut down the competitors' metabolism. For example, at pH 7, iron reducers can use H_2 when the partial pressure of hydrogen is as low as 10^{-4} atm, whereas for the hydrogenotrophic methanogens this limit is almost an order of magnitude higher [6]. Accordingly, the concentrations of hydrogen in sediments where iron reduction is the dominant respiratory process are around 0.05 nM, whereas methanogenic sediments require a buildup to around 2–12 nM [6]. Microbial respiration lowers hydrogen concentrations in sediment porewater, increasing the thermodynamic favorability for fermenting organisms that produce it. The fermenting and respiring microbes are thus in a mutually beneficial syntrophic relationship.

Test your understanding: Verify the above numbers for the hydrogenotrophic metabolisms using thermodynamic calculations at typical sedimentary conditions (see Table D.1). For solution, see page 193.

4.2.5 Biosynthesis

Catabolic energy is channeled into microbial growth, which involves the synthesis of new organic molecules for the constituents of new cells. Some compounds, such as amino acids, nucleotides, monosaccharides, and fatty acids can be assimilated directly from the environment. Others need to be made within the cell via anabolic reactions. Being the opposite of oxidative degradation, the synthesis of new biomass is a reductive process, which reduces carbon atoms and requires energy. The catabolism's energy, in addition to being stored as ATP, is used to generate the needed reducing equivalents (NADH or NADPH). Biosynthesis within the cell can be aided by the presence of molecular building blocks that are either produced or used during the catabolism: low-molecular-weight organic molecules or, to a lesser degree, catabolic products such as CO_2 (Fig. 4.5).

4.3 Microbial Growth

Synthesis of new microbial biomass can be restricted by the availability of energy sources, nutrients, or some other needed ingredients, such as sources of carbon. In environments where energy is sufficient for growth, rates of biomass synthesis are more or less proportional to the rates of respiration. In energy-starved environments, such as in the deep subsurface, cells may metabolize without producing much new biomass, with catabolic energy being expended mainly on cell maintenance. In environments where energy is abundant to the point of not being a limiting resource, biomass synthesis becomes independent of respiration and depends instead on the availability of nutrients or, if nutrients are also abundant, on the kinetic limits of cell physiology. These considerations justify the shapes of typical (Monod-type) kinetic dependencies (Fig. 4.6; see also Section 11.3.2).

Commonly used to characterize the speed and efficiency of microbial growth are parameters such as *growth rate* (μ, h^{-1}) and *growth yield* (Y, e.g., mol C biomass per mol of electron donor). The growth rate μ, having units of inverse time, effectively describes the doubling rate for the cells. As during unlimited growth the cells multiply exponentially, its inverse gives the characteristic time scale for the number of cells to increase by a factor of $e = 2.718...$, so the doubling time is $\tau_2 = \ln 2/\mu$. The growth yield Y describes the efficiency with which catabolic energy is converted into anabolism. For instance, sulfate reducers, which grow by oxidizing acetate with sulfate (SO_4^{2-}), have Y values around 0.05 molC/mol$_{ed}$, meaning that for each 100 atoms of organic carbon oxidized during catabolism, 5 carbon atoms are built into the microbial biomass. Organisms that utilize reactions with higher ΔG, such as iron reducers, would be expected to have higher yields, though not necessarily higher growth rates. The growth rate, while dependent on how much biomass is synthesized

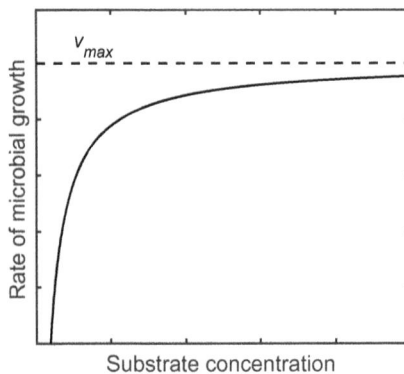

Fig. 4.6. Microbial growth dependence on substrate availability. Low concentrations of substrate slow down the kinetics of enzymatic catabolic reactions. At very low concentrations, energetic limitations restrict the use of catabolic energy to cell maintenance, stunting growth.

per reaction, also depends on how quickly the reaction is turned over by the cell: $\mu = v_{max}Y$. The cell-specific reaction rate v_{max} (mol_{ed}/h) can be highly dependent on microbial physiology and the types of substrates that the organism needs to access. Kinetics for organisms that rely on accessing mineral grains, such as iron reducers, can be substantially slower than for those that source their substrates from bulk solution, like sulfate reducers.

Most catabolic reactions have activation energy barriers (Fig. 3.7) that prevent them from happening spontaneously and abiotically. Enzymes are proteins produced by microbial cells that lower the activation energy, dramatically increasing the rates of reactions [4]. Their structure is closely tailored to the reactions they catalyze, with the stereochemistry, polarity, and charge at the active site of the enzyme molecule fitting a specific substrate. To run their complex cellular machinery, cells generate thousands of different enzymes. As catalysts, enzymes themselves are not consumed by the reactions. When the concentration of a substrate becomes too high for the number of enzyme molecules, however, the reaction rate becomes independent of the substrate concentration. This causes saturation, as described by the Michaelis–Menten equation of enzymatic kinetics (Fig. 4.6). More detailed mathematical descriptions of microbial kinetics are given in Section 11.3.2.

Effects of temperature and the pH

Each microorganism has a specific temperature range and pH range for optimal growth (Fig. 4.7). Temperature affects enzyme structures and rates of their activity, as well as the thermodynamic energy gain of the catalyzed reactions. Low temperatures decrease enzyme activity, while temperatures that are too high can denature proteins, destroying cell components. As diagenetic reactions often take place close to their thermodynamic limits (Section 11.3.2; [38]), temperature dependences may have significant effects, as energy gains and heights of activation barriers change. As the associated gains in the favorability of catabolic reactions differ among metabolisms, such changes can potentially alter the outcomes of microbial competitions, causing shifts in the relative dominance of biogeochemical pathways.

The pH has a similarly strong effect on microbial metabolisms [39]. As described above (Chapter 3), it strongly affects ambient chemical concentrations, speciations, and the energetics of catabolic

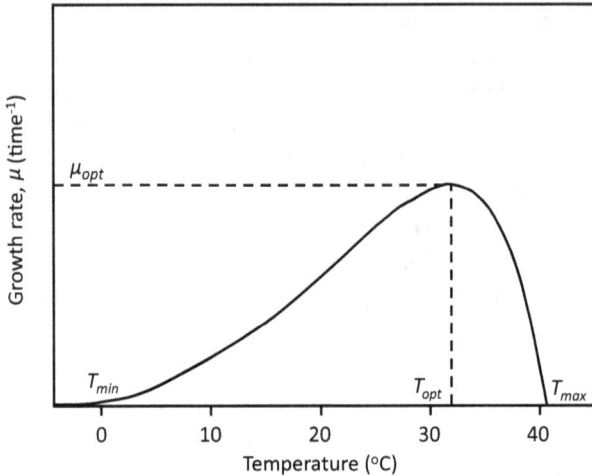

Fig. 4.7. Effects of temperature on microbial growth (adapted from [38]).

reactions [40], as well as mineral precipitation and sorption-desorption dynamics on which microbial cells may depend. pH may also affect the activities of extracellular enzymes. Changing pH by one unit can lower metabolic activity by up to 50% [41]. As a result, microbes inhabit sediment zones with pH values that closely match their preferred conditions.

4.4   Microbial Distributions within Sediments

An oft-invoked rule of thumb in microbial ecology is that for every ecological niche there is an organism to occupy it. Four billion years of evolution have shaped microbes capable of occupying nearly any environment with liquid water. Which group of organisms comes to dominate in a particular environment, however, is a complex question. Thermodynamic, kinetic, and physiological factors are all important. Additional complexity arises when one considers the competition among microbial groups for substrates and energy, and the complex networks of mutual interdependencies.

Microbes obtain energy through their catabolisms. To survive, however, the amount of chemical power they generate needs to exceed some minimum threshold that is required for the maintenance of their cells: synthesizing new enzymes, repairing DNA damage, etc.

The energy is readily available near the sediment surface but becomes progressively scarce deeper into the subsurface. The vertical sequence of microbial metabolisms thus progresses from highly energetic ones, such as aerobes, to fermenting organisms that survive near thermodynamic energy limits.

The redox structure of sediments imposes its own order on the vertical distributions of microbial communities, while also reflecting the outcomes of microbial competitions. As oxidants supplied from the sediment surface become depleted, the steadily decreasing Eh selects for reactions along the descending thermodynamic ladder (Fig. 3.4). One consequently expects to find mostly aerobic microbes in the uppermost oxidized layer, underlain by communities of nitrate reducers and manganese reducers. Descending further into the reduced sediment, one should find iron reducers, followed by sulfate reducers and finally methanogens [43]. Cohabiting with them, one should find a community of fermenting organisms, which support the respiring organisms with the low-molecular-weight products of their metabolisms. This sequence of redox zones and the associated microbial populations, however, is not universally observed in this strict order, as outcomes of microbial competitions are not shaped solely by thermodynamics [40]. The communities routinely overlap with each other and sometimes even appear in reverse order [42] (Fig. 4.8).

4.5   The Omics Methods

The tools of molecular biology – collectively nicknamed the "omics" techniques – have undergone exponentially rapid development over the past several decades. They include genomics, transcriptomics, and proteomics. The division mimics the three components of the *Central Dogma of Biology*: genes (DNA) define organisms, gene transcription rates (mRNA) relate to the speed of cell division and the rates of metabolic reactions, and proteins, which are the products of these reactions, regulate microbial capabilities in a given environment. Measurements of *functional genes*, which are the strands of DNA that can be uniquely associated with catalyzing specific metabolic reactions, reveal the abundances of organisms that are capable of such metabolisms. The rates of expression (mRNA) of such

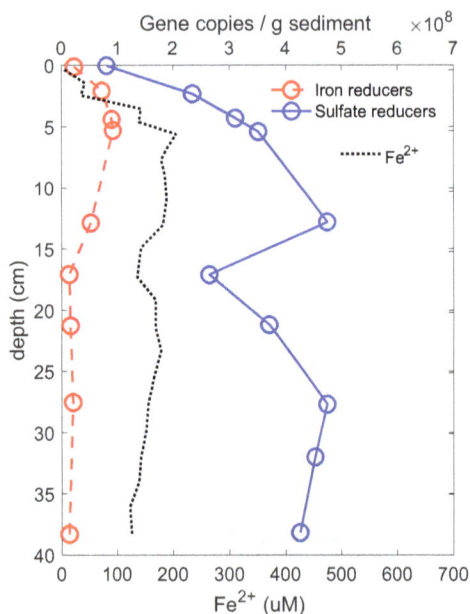

Fig. 4.8. An example of the vertical structure in a sediment microbial community, as revealed by gene copy numbers (in Cumberland Bay, redrawn from [42]).

genes characterize the rates of cell division and are associated with the levels of the organism's metabolic activities. Establishing quantitative links between the abundances of genes, their transcription rates, and the rates of the corresponding biogeochemical reactions is an ongoing challenge [44]. The field is progressing rapidly, however, driven by the flood of increasingly inexpensive data. Assisted by bioinformatics, which helps analyze the large generated datasets, techniques of molecular genomics continually improve our understanding of how elemental cycling is carried out by microbial populations, which are more numerous and more diverse than we ever realized.

4.6 Exercises

1. Using the list of common reactions in Appendix D, give an example of a reaction that might be used in catabolism by (a) a chemolithoautotroph; (b) a chemoorganoheterotroph.

2. Synthesis exercise: Figure 3.12 shows the depth profiles of several chemical species in freshwater sediment. (a) Identify how redox conditions vary with depth. (b) Write down the primary and secondary redox reactions that likely shaped these profiles, serving as sinks or sources for the reactants and products (the list in Appendix D may help). Indicate the depth intervals within the sediment where each of these reactions may be taking place. (c) Identify the types of microbial metabolisms that are likely to be involved in catalyzing these reactions.

3. Synthesis exercise: Using your knowledge and the reactions listed in Appendix D, construct a diagram that illustrates the diagenetic cycling of the following chemical constituents: (a) O_2, (b) CH_4, and (c) inorganic nitrogen compounds. Identify the likely reactions and the types of microbial metabolisms involved. Even though many of the reactions are covered only in the later chapters, you should be able to come up with reasonable predictions.

4. A *FeAmmox* reaction has been suggested as a potential microbial catabolism [45]. The reaction couples oxidation of ammonium (NH_4^+) with the reduction of iron oxides. The iron oxides are reduced to Fe^{2+}. Assuming goethite (FeOOH) as the oxidized iron form, which of the following chemical species could be produced by the reaction under relevant environmental conditions: NO_3^-, NO_2^-, N_2, N_2O?

Chapter 5

The Carbon Cycle

Carbon is the main element of Life. Being a key component of all living organisms, it is also present as carbon dioxide in air and water and exists as carbonate minerals on land and in the sea. Carbon connects the biosphere with Earth's atmosphere, hydrosphere, and lithosphere, and regulates the planet's climate system. Its cycling through ecosystems is thus central to many branches of Earth science. Small imbalances in the global cycling of carbon to lead to accumulation of greenhouse gases in the atmosphere and climate change. Too much dissolved carbon dioxide in water leads to acidification, which harms aquatic organisms. Over geological time scales, variations in the rates of carbon burial into the seafloor regulate oxygen levels in the atmosphere [7]. Sediments are key to all these processes. While only a small fraction of organic material produced by photosynthesis reaches sediments, recycling and mineralization of organic carbon within sediments affects the amount of carbon cycling through the entire ecosystem. Diagenetic cycling of carbon also determines the structures of microbial communities, fluxes of nutrients into the water column, and the amounts of greenhouse gases that escape into the atmosphere. Reactions that occur in the upper centimeters of sediments determine the chemical properties of the entire aquatic environment and the conditions in benthic habitats. In this chapter, after briefly outlining the broader context of diagenetic carbon cycling, we analyze the biogeochemical processes that govern the decomposition of organic matter the recycling of its products, and that determine the fraction of organic material that becomes permanently buried.

5.1 Organic Carbon Mineralization

Carbon compounds in aquatic ecosystems are commonly subdivided into two types: organic and inorganic. Organic carbon comes from living and dead organic matter, including its decomposition products. It is operationally divided into particulate organic carbon (POC) and dissolved organic carbon (DOC), based on whether it passes through a fine filter (usually 0.2 or 0.7 micron). The DOC fraction consists primarily of the decomposition products that are released during the process of POC mineralization. Inorganic carbon includes aqueous species of carbon dioxide (CO_2, HCO_3^-, and CO_3^{2-}, collectively termed the *dissolved inorganic carbon*, DIC), solid minerals that form from them, such as calcium carbonate, and methane (CH_4).

5.1.1 Hydrolysis

As the deposited organic matter becomes buried into sediment, the organic molecules undergo a complex process of breakdown. Particulate organic matter (POM) becomes hydrolyzed, as microbial enzymes cleave chemical bonds, transferring organic molecules into the solution. Some of the resultant dissolved organic matter (DOM) is not reactive and can escape further mineralization for long periods of time. In fact, DOM dominates the detritus pool in the water column, accounting for >95% of the total organic matter in oceans,

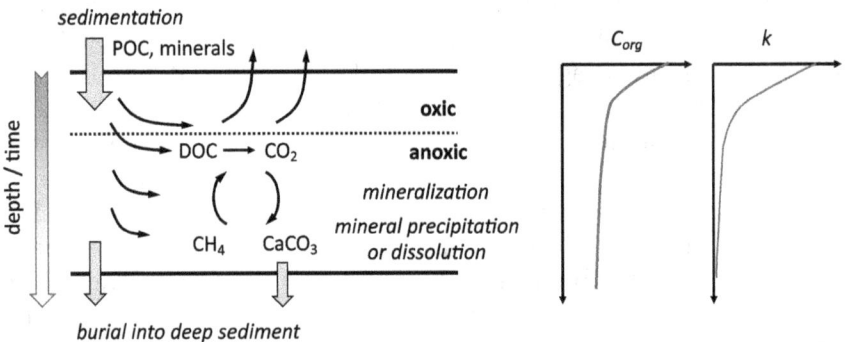

Fig. 5.1. Fluxes of carbon through sediments. Graphs on the right illustrate decreases in the concentration (C_{org}) and reactivity (k) of organic carbon.

lakes, and rivers [6]. Due to its production during diagenesis, the concentrations of dissolved organic carbon (DOC) in sediment porewater reach mM levels or higher, an order of magnitude greater than in the water column. Sediments therefore continually leak DOC into the water column. A significant proportion of the DOC, however, continues to be degraded within the sediment, fueling a crucial sequence of diagenetic redox reactions (Section 5.1.3).

5.1.2   Types of organic matter compounds

Organic matter that settles into sediment consists of the remains of aquatic organisms and the solid products of their metabolisms (feces). Despite its chemical diversity, it can be broken down into a relatively small number of groups of compounds [46].

Proteins are long molecules built from *amino acids*. Proteins are easily degradable, and the concentrations of amino acids in the POM fraction usually decrease with depth into the sediment. *Nucleic acids*, which include the DNA and RNA molecules of all living cells, are similarly readily broken down and hydrolyzed during diagenesis. Due to their nitrogen-containing base units, nucleic acids account for a substantial portion of the N-bearing compounds in cells, which also makes them an important source of nitrogen after mineralization.

Carbohydrates make up a significant proportion of the biomass, especially in plants, and include several groups, such as saccharides, sugars, and others. *Polysaccharides* make up structural tissues (cellulose, chitin) and energy storage compounds, such as starch. While insoluble, many of them can be quickly hydrolyzed during diagenesis to soluble sugars, which are highly bioavailable. *Lignin*, another structural component of vascular plants, on the other hand, is highly refractory and can remain undegraded for a long time.

Lipids are insoluble in water and include a wide range of compounds, such as saturated and unsaturated fatty acids, steroids, carotenoids (such as plant pigments), and chlrolophylls. Their resistance to hydrolysis makes them less likely to decompose quickly and more likely to be preserved in sediment organic matter.

Different groups of aquatic and terrestrial organisms whose remains make up sedimentary organic matter contain these types of compounds in widely different proportions (Table 5.1). And differences in the ability of these compounds to undergo hydrolysis and

Table 5.1. Composition of organic matter endmembers (after [3, 46]). See also Table D.1.

Organic matter type	Proteins (%)	Carbohydrates (%)	Lipids (%)	Lignin (%)
Marine organic matter	50–60	20–40	5–30	0
Phytoplankton	30	20	5	0
Phytobenthos	15	60	0.5	0
Zooplankton	60	15	15	0
Zoobenthos	27	8	3	0
Vascular plants	1–2	≈70	1–2	≈30

mineralization lead to differences in their degrees of degradation during sediment diagenesis. In marine sediments, estimates suggest that about 33–47% of amino acids become remineralized during diagenesis, versus 24–55% for sugars and 8–33% for lipids [3].

Compared to particulate organic matter, dissolved organic matter is much less defined in its composition. Only a minority of DOM molecules have been grouped into major classes such as carbohydrates, amino acids, and lipids [6]. The characterized fraction is usually dominated by carbohydrates. Humic substances may account for a significant portion of DOM in freshwater lakes, but only about 15–25% in marine environments.

5.1.3 The redox cascade

Decomposition of organic carbon is conducted in sediments by a diverse and interconnected community of microorganisms, whose chemotrophic metabolisms are fueled by carbon oxidation. In a chemical sense, mineralization is photosynthesis in reverse. Photosynthesis reduces the carbon atoms of CO_2, storing the energy of sunlight as the chemical energy of complex organic molecules. Mineralization releases this chemical energy, reoxidizing the carbon within organic matter, mostly back to carbon dioxide.

The oxidation reaction can take different reduction reactions as its redox partners. When oxygen is available, it is the oxidant of choice. Just as we breathe oxygen to oxidize the carbon in our food to CO_2, the microbial consortia in oxic sediments conduct the overall reaction of aerobic respiration:

$$C_{org} + O_2 \longrightarrow CO_2 \qquad (5.1)$$

The generic notation for organic carbon, C_{org}, here represents a complex array of organic compounds. For simplicity, the stoichiometry of carbon compounds within organic matter is often approximated by the stoichiometry of glucose: $C_6H_{12}O_6$, or CH_2O when expressed per carbon atom. When the nutrients nitrogen and phosphorus are taken into account, the same reaction of the aerobic mineralization of natural organic matter can be written as

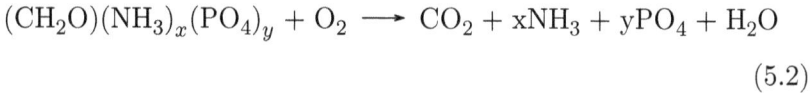

$$(CH_2O)(NH_3)_x(PO_4)_y + O_2 \longrightarrow CO_2 + xNH_3 + yPO_4 + H_2O$$

$$(5.2)$$

For typical marine organic matter, the stoichiometric coefficients x and y are well approximated by the Redfield C:N:P ratios of 106:16:1, meaning $x = 1/16, y = 1/106$. In freshwater, these ratios can be substantially more variable [47].

Unlike higher organisms, which all have to breathe oxygen, microbes can respire other electron acceptors. As long as there is an energy gain, the combined redox reaction can be used to power microbial catabolism. The most significant electron acceptors that replace oxygen under anoxic conditions are nitrate (NO_3^-), manganese oxides (MnO_2), iron oxyhydroxides (nominally $FeOOH$ or $Fe(OH)_3$), and sulfate (SO_4^{2-}) (Table 5.2). Being relatively abundant in the environment, they account for a significant proportion of sediment respiration. As discussed below, the reduced products of the corresponding redox reactions have their own significance: they enable a network of diagenetic reactions that connect the biogeochemical cycles of nitrogen, iron, manganese, sulfur, and the associated elements such as phosphorus and trace metals.

The extent to which the oxidation of organic carbon uses each of these pathways depends on several factors, such as the availability of each oxidant at a given location within the sediment and the physiological capabilities of the microbes that determine how fast they can turn over the corresponding reactions. The primary organizing principle, however, is commonly accepted to be related to the amounts of energy that each of these competing pathways provides.

Table 5.2. Organic carbon mineralization sequence. Organic compounds are represented by a generic formula CH_2O, which is stoichiometrically equivalent to $\frac{1}{6}C_6H_{12}O_6$ (glucose). ΔG^0 values were calculated at pH 7 per mol of organic carbon (kJ/molC), for the reactions as written. ΔG values correspond to pH 7, $T = 4°C$, using glucose as organic carbon source, goethite as FeOOH, pyrolusite as MnO_2, $[HCO_3^-] = 1$ mM, and concentrations of all other dissolve species at 10 μM. See also Table D.1.

Reaction	ΔG^0 (pH 7)	ΔG
Oxic respiration		
$CH_2O + O_2 \longrightarrow HCO_3^- + H^+$	-488	-451
Nitrate reduction		
$CH_2O + \frac{4}{5}NO_3^- \longrightarrow HCO_3^- + \frac{2}{5}N_2 + \frac{2}{5}H_2O + \frac{1}{5}H^+$	-448	-417
Manganese reduction		
$CH_2O + 2\,MnO_2 + 3\,H^+ \longrightarrow HCO_3^- + 2\,Mn^{2+} + 2\,H_2O$	-324	-366
Iron reduction		
$CH_2O + 4\,FeOOH + 7\,H^+ \longrightarrow HCO_3^- + 4\,Fe^{2+} + 6\,H_2O$	-117	-212
Sulfate reduction		
$CH_2O + \frac{1}{2}SO_4^{2-} \longrightarrow HCO_3^- + \frac{1}{2}H_2S$	-76	-65
Methanogenesis		
$CH_2O + \frac{1}{2}H_2O \longrightarrow \frac{1}{2}CH_4 + \frac{1}{2}HCO_3^- + \frac{1}{2}H^+$	-58	-53

This principle underlies the concept of the *redox cascade*, or *thermodynamic ladder*: electron acceptors that provide the highest amounts of Gibbs free energy (Table 5.2 and Fig. 3.4) are likely be to the ones most beneficial to the microbial communities that catalyze them. Accordingly, they are going to be used preferentially, before the less energetic electron acceptors.

A few millimeters or centimeters below the sediment surface, where oxygen becomes depleted to a few μM, nitrate respiration begins (Fig. 5.2). This is the zone of nitrate reduction, or *denitrification*. A bit deeper, where nitrate is depleted, manganese oxides are used. Next, iron reduction occurs below the depth where manganese oxides are depleted, and sulfate reduction becomes efficient after the reactive iron oxides are exhausted. When all externally supplied electron acceptors are depleted, further mineralization of organic carbon

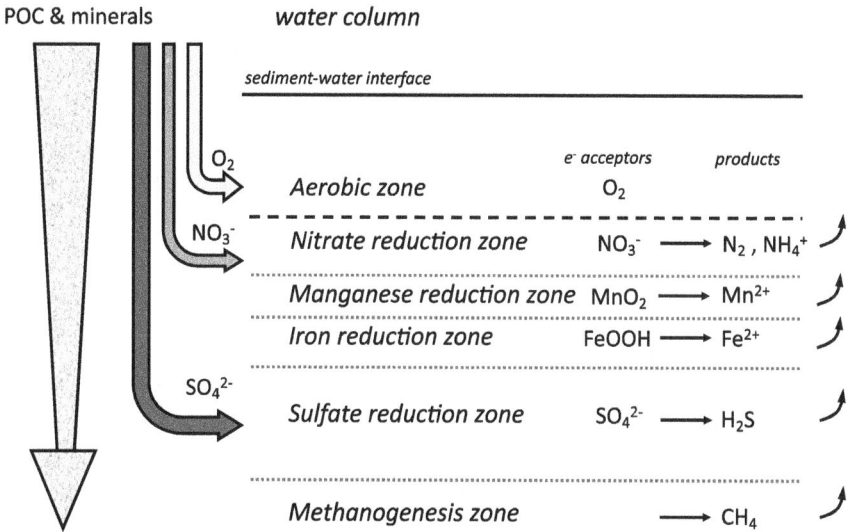

Fig. 5.2. Diagram of the redox zonation that results from the thermodynamic ladder of microbial metabolisms in the organic carbon mineralization sequence.

proceeds via fermentation, which uses organic molecules as both electron donors and electron acceptors. Methanogenic fermentation (carried out exclusively by archaea) transfers electrons between the carbon atoms within organic molecules such as acetate, producing both oxidized (CO_2, +4) and reduced (CH_4, −4) inorganic carbon (Table 5.2). Generating less energy than the respiratory pathways, methanogenesis is the last stage of early-diagenetic degradation of organic matter. It may continue, at progressively diminishing rates, many meters deep into the sediment. While thermodynamics is not the only controlling factor and these processes do not necessarily occur in all sediments in this precise sequence, this thermodynamic framework [48] serves as a highly useful guide to the key processes that one typically finds in a downward sequence in sediments.

The justification for the thermodynamically-dictated order of reactions is rooted in microbial ecology. With energy being a significant contributor to an organism's fitness, organisms with access to more energy can multiply faster and use key substrates at faster rates than their competitors. Other considerations, such as toxicity, add complexity and nuance. For example, as oxygen is toxic to most anaerobes, only nitrate reduction is possible at any significant rate in the oxic sediment zone, and only at oxygen concentrations that do not exceed several 10s of μM. Organisms also differ in the types of organic substrates they can access. As discussed in Chapter 4, most anaerobic respiring microbes can access only relatively simple, small organic molecules and require assistance from a plethora of (anaerobic) fermenting organisms that hydrolyze and break down the more complex organic compounds (Fig. 5.3). Aerobic microorganisms (and to some degree nitrate reducers), on the other hand, can access the particulate organic matter more directly.

Test your understanding: Can you suggest an example of a situation where the order of the redox reactions in Fig. 5.2 would be different? Consider scenarios when this might happen for a thermodynamic reason (see e.g., Fig. 3.4), a kinetic or other chemical reason, or a biological reason.

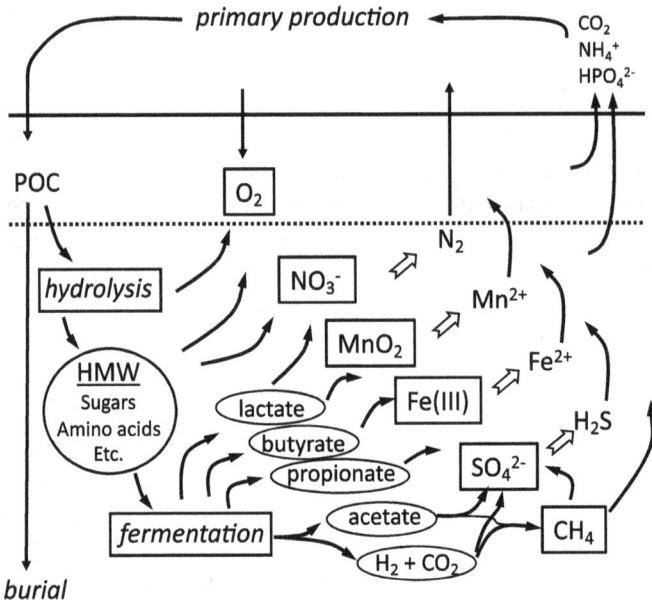

Fig. 5.3. The substrate cascade within the redox cascade of microbial metabolisms. Different metabolisms rely on different organic substrates. Organisms that consume external electron acceptors for their anaerobic respiration rely on a community of fermenters for their supply of low-molecular-weight organic electron donors (redrawn from [37]).

5.1.4 Vertical redox zonation

The redox cascade of organic carbon mineralization creates a vertical progression of distinct redox zones within the sediment (Fig. 5.2). Each zone is characterized by a specific set of chemical reactions (Fig. 5.2) and a distinct chemical composition that results from them (Fig. 5.4). Each zone also hosts a specific set of microbial communities (Fig. 4.8).

As long as the water above the sediment contains oxygen, the uppermost layer of the sediment is oxidized. In most lakes and muddy coastal marine sediments, the thickness of this layer does not exceed several millimeters. In well-oxygenated and organic-poor environments, such as the deep ocean or some of the Great Lakes [11], oxygen penetrates several cm or even more [49, 50]. The oxic layer hosts aerobic microbial communities that oxidize both organic carbon and the reduced products of anaerobic metabolisms from deeper sediment layers. The high oxidizing potential of oxygen keeps most of the redox-sensitive elements within the oxic layer in their oxidized states. Reduced substances that enter this layer are quickly oxidized. Reduced nitrogen species, such as ammonium generated during the decomposition of organic matter, are oxidized to nitrate

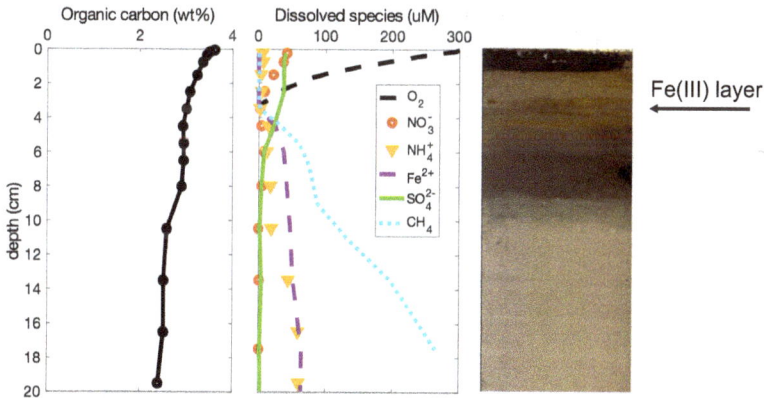

Fig. 5.4. Redox zonation in the sediments of Lake Superior. Distinct zones are seen in the color of the sediment and in the distributions of oxidants and reduced products. Organic carbon content decreases with depth into the sediment. Data are from [51].

(Chapter 6). Iron and manganese minerals remain in their insoluble oxide forms (Chapter 7). The more abundant iron oxides often give reddish color to the oxidized sediment (Fig. 5.4). Oxidation also keeps the surface layer free of hydrogen sulfide (H_2S), which is toxic to benthic animals.

The maximum depth of oxygen penetration marks the most significant chemical boundary within the sediment. As redox-sensitive elements are oxidized above this boundary and reduced below it, their cycling across this boundary recycles them within the sediment, creates a hotspot for chemical reactions, and generates distinct layers of minerals at the boundary. Diagenetic enrichments of black manganese oxides and red iron oxides continually form near the oxic-anoxic transition (see Chapter 7) and are sometimes visible to the naked eye (Fig. 5.4). Dissolved chemical species, such as ammonium or methane that diffuse upward from the reduced sediment are oxidized at the boundary, which limits their fluxes into the water column.

In the anoxic sediment below, reduced products of anaerobic metabolisms accumulate. The reduced forms of manganese and iron, Mn^{2+} and Fe^{2+}, are soluble in water. The reduction of solid Mn and Fe oxides thus mobilizes these elements into the porewater, and they can migrate vertically within the sediment. Reduced porewaters similarly accumulate the dissolved products of other anaerobic metabolisms: hydrogen sulfide (H_2S), methane (CH_4), as well as ammonium (NH_4^+) that results from the breakdown of organic molecules by all metabolisms (Eq. (5.2)). These aqueous species can diffuse upward, where they can be reoxidized by oxygen or by another electron acceptor. Such *secondary redox reactions* are important for the recycling of elements within the sediment (see Chapters 6, 7, and 9). The nitrogen gas (N_2) produced by denitrification is not reactive and usually escapes from sediment.

5.1.5   Effects on the pH

Both the mineralization of organic carbon and the secondary redox reactions involve consumption or production of hydrogen ions (Table 5.2), meaning that they may affect the pH. At circumneutral pH, the oxic mineralization of organic matter produces hydrogen ions, decreasing the pH (the opposite of photosynthesis, which increases the pH). In reduced sediment, metal reduction consumes protons, while denitrification (nitrate reduction), sulfate reduction, methanogenesis, and secondary reoxidation reactions generally produce them (Table D.1). As a result, vertical profiles of the pH in marine sediments often exhibit a minimum at the bottom of the oxidized layer, followed by a small peak around the zone of metal reduction, followed by a slow decrease in pH into deeper sediment (for more details, see e.g., [52,53]). Variations that run contrary to such trends could sometimes identify unusual processes, such as the activities of giant *cable bacteria*, which transport electrons vertically over distances of several centimeters from hydrogen sulfide to oxygen [54].

5.1.6   Oxygen uptake

Despite the range of electron acceptors used by organic carbon mineralization, the overall rate of mineralization in the entire sediment column can be fairly accurately estimated from the rate of oxygen consumption. This is because most of the reduced species generated by anaerobic metabolisms (Mn^{2+}, Fe^{2+}, H_2S/HS^-, CH_4) are reoxidized within the sediment (Table D.1), with oxygen serving as the ultimate eventual oxidant. The additional demand for oxygen created by the secondary reoxidation reactions makes oxygen consumption stoichiometrically equivalent to the oxidization of organic carbon. For instance, if four atoms of iron are needed to oxidize one atom of carbon (Table D.1), then subsequent reoxidation of those four atoms of iron by oxygen would consume one molecule of O_2 (Table D.1), resulting in a 1:1 carbon-to-oxygen consumption ratio. Exceptions are denitrification, which produces the unreactive N_2 gas that leaves the sediment (Table 5.2), and co-precipitation of Fe^{2+} with H_2S (Eq. (7.9)), which traps both species in the reduced sediment. Additional oxygen demand is placed by the oxidation of ammonium (NH_4^+) that is generated during the organic matter degradation. For a typical Redfield C:N stoichiometry, it can account for about $x = 1/16 = 6.25\%$ of the total oxygen demand. This approximately offsets the effect of denitrification, which typically accounts for a similar percentage of the overall carbon mineralization (Section 6.2.3). The total oxygen consumption for the sediment (*total oxygen uptake* (TOU), or total sediment oxygen demand (SOD)), in many cases, is thus approximately equivalent, within a few percent, to the total rate of carbon mineralization. Measuring the sediment oxygen uptake is therefore a useful and often easier technique to quantify the mineralization of carbon and the overall production of CO_2. Such measurements can be performed *in situ*, in laboratory incubations of sediment cores, or by analyzing the vertical profiles of dissolved oxygen in sediment porewaters (see Appendix A and Exercise 4). Collecting fine-scale concentration profiles of oxygen with profiling microsensors has become a standard technique that can additionally provide a wealth of quantitative information about the vertical variations in the rates of carbon mineralization, and even the intensities of bioturbation and bioirrigation (see Chapter 11).

Test your understanding: The flux of oxygen into a sediment at steady state is 5 mmol $O_2 \, m^{-2} \, day^{-1}$. In the deep sediment, the burial flux of organic carbon is estimated at 2 mmolC $m^{-2} \, day^{-1}$. What is the sedimentation flux of organic carbon to the sediment surface?

5.1.7   Methanogenesis and methanotrophy

Methane (CH_4) is produced within sediments by two pathways: using either acetate (CH_3COO^-) or molecular hydrogen (H_2) as electron donors.

Acetoclastic methanogenesis, is a non-respiratory pathway of organic carbon mineralization (Table 5.2):

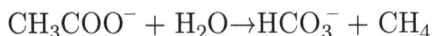

$$CH_3COO^- + H_2O \rightarrow HCO_3^- + CH_4$$

Hydrogenotrophic methanogenesis, in contrast, does not require organic molecules as electron donors and instead uses molecular hydrogen to reduce CO_2:

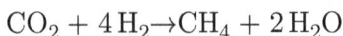

$$CO_2 + 4\,H_2 \rightarrow CH_4 + 2\,H_2O$$

Not relying on organic substrates allows hydrogenotrophic organisms to be autotrophs, i.e. to use ambient CO_2 also as the carbon source for their biomass. While this pathway does not participate in the mineralization of deposited organic carbon, except by utilizing H_2 produced by fermentation, it nevertheless can be an important part of the overall cycling of carbon, as it partitions carbon between carbon dioxide and methane.

Methane is oxidized within sediments by several electron accep-tors (Fig. 3.4). The most common oxidants are oxygen and sulfate (Table D.1). Aerobic oxidation by oxygen happens when methane reaches the uppermost oxidized layer. The *anaerobic oxidation of methane* (AOM) by sulfate happens deeper in the sediment. In marine environments where sulfate is abundant, the oxidation of methane at the bottom of the sulfate reduction zone – in a region sometimes termed the sulfate-methane transition zone (SMTZ) – is an important process that prevents vast quantities of produced methane from reaching the water column and ultimately the atmo-sphere. Freshwater AOM has been reported [55] but is unlikely to be of widespread significance, given the much lower levels of sul-fate than in seawater. Oxidation of methane by iron and manganese oxides [56,57], as well as by nitrate [58,59], has also been reported. Potentially significant in niche environments with high availability of these electron acceptors, so far these pathways have been rarely con-sidered to be of broad significance for methane cycling. Together, aerobic and anaerobic oxidation processes prevent methane from reaching the sediment surface and the water column. They may be bypassed, however, when active methanogeneses leads to the for-mation of gas bubbles within the sediment. Methane ebullition (see Section 5.1.8) can be a significant alternative mechanism of methane emissions from water bodies.

5.1.8 Gas ebullition

Due to the relatively low solubility of methane in water (26 times lower than that of carbon dioxide), high rates of methanogene-sis in organic-rich sediments may cause methane bubbles to form within the sediment. These Bubbles may then grow, creating frac-tures in the sediment matrix that can serve as conduits for subse-quent bubbles, and can escape into the overlying water [60,61]. As they rise, the decreasing hydrostatic pressure causes the bubbles to expand, which may lead to their breakup. Bubble plumes are some-times observed rising from sediments, creating distinct signatures on echosound images [62]. In deep lakes, low concentrations of methane in the water column may cause methane bubbles to redissolve on their way up.

The ebullition of methane from sediments is one of the major modes of methane emission from lakes [63] and an underappreciated source of this greenhouse gas into the atmosphere. Freshwaters contribute about 20% of all global natural CH_4 emissions [64], and about half of those fluxes are attributed to ebullition [65]. Ebullition is particularly prevalent in reservoirs, where sediments are organic-rich. Bubbles can grow and escape sediments especially rapidly when the water level in a reservoir fluctuates, temporarily decreasing the hydrostatic pressure [66].

5.2   Carbon Fluxes Through Sediments

5.2.1   Sedimentation

Carbon enters sediments primarily with particles settling from the water column. The organic material produced in the sunlit upper waters decomposes on its way down, so only a fraction of it reaches the sediment [67], even in shallow lakes (Fig. 5.5).

The concentrations of organic carbon in sediments vary widely, especially in freshwater environments. Pelagic marine sediments tend to have TOC content below 0.5 wt%. Coastal areas range typically up

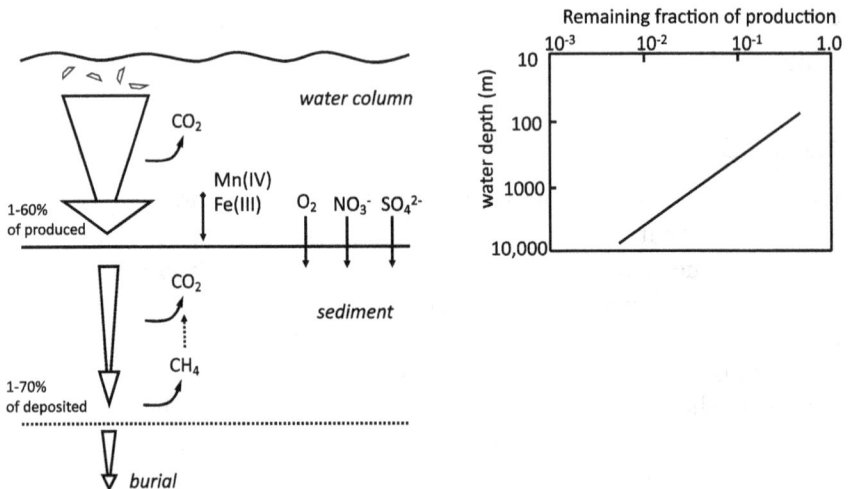

Fig. 5.5. Carbon cycling through aquatic systems. In deep water bodies, only a small percentage of produced organic matter that settles through the water column reaches the sediment. Only a fraction of that deposited amount becomes buried. The approximately power-law decrease in the remaining fraction of organic matter as a function of depth in the water column is redrawn from [67].

to about 3 wt%, and up to 6 wt% in highly productive areas [15]. In contrast, "organic-poor" muddy sediments even of large oligotrophic lakes have TOC content around 3–6 wt% [11,68], and sediments of small productive lakes can contain 15 wt% or even more.

Organic compounds continue to be broken down in the sediment, so only a small fraction of the deposited carbon is ultimately buried into the deep sediment. In the process, some of the externally brought organic compounds are used as building material for the new biomass of the sedimentary microbes and macrofauna, which in turn also eventually become organic debris. Recycling of organic matter plays an important role in the functioning of ecosystems, as it supplies CO_2, methane, and nutrients to the water column.

5.2.2 Mineralization rates

Organic carbon mineralization ultimately fuels most of the geochemical reactions of sediment diagenesis, so the rates of mineralization and the amounts of reactive organic matter remaining become important kinetic controls. As labile fractions of organic material become degraded before refractory ones, the overall reactivity of organic matter decreases with time after deposition. Calculating the amount of organic carbon that remains after a given time is a complicated problem and is unlikely to have a simple answer [69,70]. The rates of breakdown have been variably described as being affected by factors such as the presence of oxygen, properties of the mineral matrix, sulfurization of organic matter due to the presence of hydrogen sulfide (euxinia), temporal variations in sediment redox conditions, physical reworking of sediment by benthic animals and bottom currents, and so on. Despite of this complexity, simple models of organic matter mineralization have been widely used to obtain practically useful approximations (see Chapter 11). An approximate relationship that seems to hold in both marine [71] and freshwater [72] sediments is that the reactivity of the bulk organic matter appears to decrease with its age according to a power law. When the decomposition rate of organic carbon is described by first-order kinetics

$$R = kC \tag{5.3}$$

the proportionality constant k between the rate R and the concentration C decreases with the age of organic material t by following a power-law function of time:

$$k(t) = at^{-b} \tag{5.4}$$

As the organic material transitions from the oxic into the anoxic part of the sediment, the mineralization rate seems to decrease by about a factor of 2 (evidenced by the vertical shift in the corresponding regression lines in Fig. 5.6). This effect of oxygen is well known and agrees with the current understanding of the differences between heterotrophic microbial metabolisms in oxic vs anoxic conditions (Chapter 4). In contrast, below the depth of oxygen penetration, the rates of carbon mineralization do not seem to change between the zones of different anaerobic metabolisms, despite the obvious decreases in the available energy down the redox sequence. For example, no appreciable change in the overall mineralization rate is observed between the zones of sulfate reduction and methanogenesis [73]. A plausible hypothesis for this insensitivity is that the kinetics of the overall mineralization reactions is limited not by the final step that involves the terminal electron acceptor, but by the speed

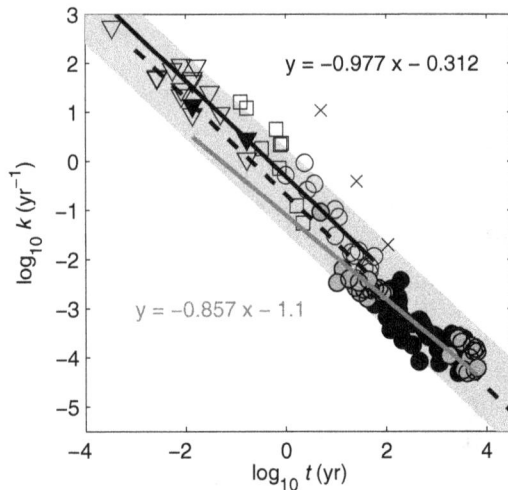

Fig. 5.6. The phenomenological power law of organic matter mineralization. Solid lines are the linear regression fits to the data in oxic (black) and anoxic (grey) freshwater environments [72]. The dashed line is the overall power law reported in marine environments [71]. (Figure from [72]; copyright ©The Geological Society of America, used with permission.)

of the enzymatic hydrolysis of organic particles, which depends on the properties of the organic material itself.

5.2.3   Preservation and burial

The processes of organic carbon mineralization determine the fraction of the deposited carbon that becomes permanently buried into the deep sediment, rather than recycled (e.g., as CO_2, CH_4, or DOM) back into the water column. The efficiency of carbon mineralization (or, alternatively, the efficiency of carbon burial) depends on the conditions within the sediment. The type of organic matter that gets deposited into the sediment, the intensity of physical reworking by bioturbation and sediment resuspension, the mineral composition of the inorganic sediment matrix, and the concentrations of oxygen and hydrogen sulfide have all been implicated as factors affecting the rates of mineralization and preservation. Organic material is preserved better in sediments when it is less exposed to oxygen (Fig. 5.7), when it is sulfurized through exposure to hydrogen sulfide, and when it is not resuspended or reworked by benthic fauna. A high proportion of terrestrial organic matter, which tends to be more refractory, also leads to a higher fraction remaining unmineralized. The amount of carbon that reaches the sediment decreases with water depth, and the material becomes less reactive as it travels through the water column (Fig. 5.7).

The difference between the rates of mineralization in different environments results in a significant effect on the fraction of the deposited organic carbon that becomes buried in different types of sediments. Anoxic, and especially euxinic (sulfidic), sediments are known to bury a much greater fraction of the deposited material than oxygenated sediments. Sediments in regions of extensive resuspension, such as river deltas, are known to be "incinerators" of organic carbon, as organic material is reworked and exposed to oxygen multiple times [75]. Figure 5.7 shows this effect on *burial efficiency*, which is defined here as the fraction of organic material deposited at the sediment surface that becomes buried permanently into the deep sediment.

The role of sediments in the overall carbon budgets of water bodies varies in a wide range (Fig. 5.8). Sediments of well-oxygenated pristine lakes incinerate their organic matter rather effectively.

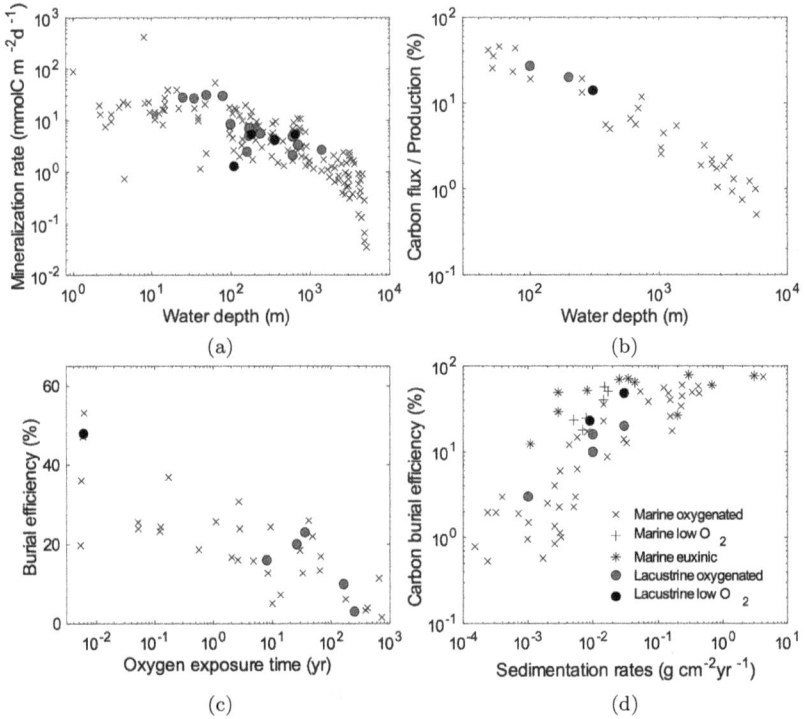

Fig. 5.7. (a, b) Organic carbon mineralization rate and the remaining fraction of primary production, as a function of water depth in marine and freshwater environments. (c, d) Organic carbon burial efficiency in freshwater and marine sediments and its relationships to the duration of oxygen exposure and sedimentation rates (adapted from [74], with permission from Elsevier).

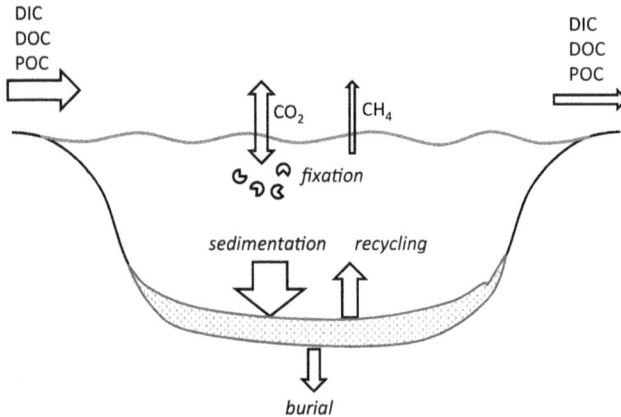

Fig. 5.8. Carbon fluxes through terrestrial aquatic ecosystems.

In contrast, organic-rich sediments that become anoxic quickly have lower rates of carbon mineralization and bury a proportionally greater fraction of the deposited organic carbon. Sediments of agricultural reservoirs, in particular, become major sites of carbon storage [76].

5.3 Exercises

1. Referring to the vertical profiles of the chemical species shown in Fig. 5.4, identify the active microbial metabolisms and the vertical zones within the sediment where they operate.
2. The total oxygen uptake in a sediment at steady state is 0.2 $mmolO_2/cm^2/y$. In the deep sediment, the concentration of organic carbon is 1 $mmolC/cm^3$, and it is being buried with a downward velocity of 0.1 cm/y. Estimate the sedimentation flux of organic carbon to the sediment surface and the efficiency of the organic carbon burial into the deep sediment.
3. The total oxygen uptake by the sediment is 5.0 $mmol/m^2/d$. The burial velocity at the bottom of the diagenetically active zone is 1 mm/y, the porosity there is 0.8, and the carbon content of the solid sediment is 2 wt%. Estimate the burial efficiency of organic carbon in this sediment, relative to the depositional flux of organic carbon from the water column. Assume steady state diagenesis.
4. The concentration of oxygen at the sediment-water interface is 100 μM and it decreases approximately linearly to zero at a depth of 0.2 cm below the interface. The porosity is 0.9. Estimate the depth-integrated rate of organic carbon mineralization in this sediment.
5. Synthesis exercise: Sketch the likely carbon cycle in: (a) a deeply oxygenated organic-poor deep-sea sediment; (b) an organic-rich sediment in a eutrophic coastal ocean area. Indicate the order of magnitude of the relevant fluxes of elements through the sediment.
6. Synthesis exercise: Compare the relative roles of sediments in terms of their contributions to the overall carbon cycling in the ecosystem and the vertical fluxes of carbon throughout the sediment, water column, and the atmosphere, for different

environments: (a) a small, shallow lake in an agricultural area; (b) a coastal ocean region; (c) an abyssal ocean region. Where possible, make quantitative estimates, both for the absolute values of carbon fluxes and their relative values, e.g., relative to the primary productivity in the corresponding ecosystem.

Chapter 6

The Nitrogen Cycle

Nitrogen (N) cycling is key to the availability of this important nutrient to organisms in many water bodies, including the global ocean. While the Earth's atmosphere is 78% nitrogen gas (N_2), only a tiny minority of aquatic organisms can use it in that form. Sediments provide the environment needed for key reactions that transform nitrogen across a spectrum of redox states. Diagenetic processes determine the kinds of nitrogen compounds that become recycled into the water column in biologically available forms, and the fraction of nitrogen that becomes removed from active circulation. This chapter deconstructs the cycle of nitrogen by going over its individual components – chemical species and reactions. It then places them together in a vertical sequence in the sediment. After reviewing the relevant chemical species and reaction pathways, we analyze how the interplay of these processes determines the benthic sources and sinks for reactive nitrogen.

6.1   Reactive Nitrogen Species

Nitrogen (N) is a major component of all living matter and is actively exchanged by organisms with their environment. While more abundant than phosphorus (Chapter 8), it is still a productivity-limiting nutrient in large swaths of the ocean and in many lakes, as organisms require 12–16 times more N than P. Not all forms of nitrogen, however, are equally useful to organisms. Some of the important transformations that affect biological availability occur

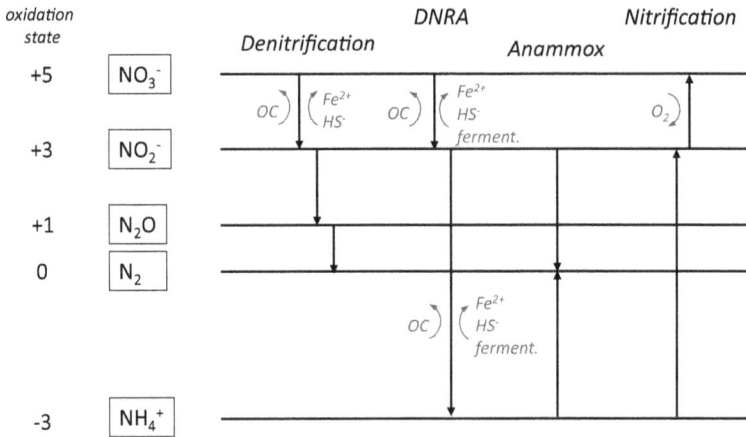

Fig. 6.1. Biogeochemical transformations involving common nitrogen compounds: nitrate (NO_3^-), nitrite (NO_2^-), nitrous oxide (N_2O), nitrogen gas (N_2), and ammonium (NH_4^+). DNRA = dissimilatory nitrate reduction to ammonium.

only in environments devoid of oxygen. For that reason, the availability of nitrogen to aquatic organisms throughout the water column is strongly affected by the processes that take place in sediments.

Nitrogen has multiple redox states (Fig. 6.1). The oxidative and reductive transitions between them occur in opposite directions in the uppermost (oxygenated) and deeper (anoxic) sediment zones. Together, they create a complex network of reactions that fuel the catabolisms of diverse microbial populations. Organisms also assimilate N (mostly as ammonium and nitrate) from the environment to satisfy their nutrient requirements.

The two most abundant chemical species of N are nitrate (NO_3^-) and ammonium (NH_4^+). They correspond, respectively, to nitrogen's most oxidized ($+5$) and most reduced (-3) states. Ammonium, which is the ionic form of ammonia (NH_3), dominates the overall NH_3/NH_4^+ pool at circumneutral as well as acidic pH (Fig. C.2). As ammonium is produced during the decomposition of organic matter, its concentrations in the reduced sediment porewater may reach upwards of several hundreds of micromolar. In contrast, the concentrations of nitrate in the oxic waters overlying the sediment are typically on the order of 10s of micromolar. The intermediate oxidized species, nitrite (NO_2^-, +3), is less abundant, typically at μM levels. In sediments, it is typically found only within narrow depth intervals but serves as an important reaction intermediate and cycles rapidly. Even less abundant in aquatic environments are the intermediate nitrogen oxides NO (+2) and N_2O (+1). While the concentrations of N_2O are typically sub-micromolar, being a dissolved gas rather than an ion makes it easier for it to escape from sediment. Despite its high solubility (Table C.4), fluxes of N_2O from sediments are environmentally important, as it is a potent greenhouse gas.

Organic nitrogen is contained in proteins (including amino acids), nucleic acids (5–10% of the living soft biomass), and many other key molecules. Operationally, it is commonly divided into dissolved organic nitrogen (DON) and particulate organic nitrogen (PON), similarly to the DOC and POC fractions for organic carbon. Though microorganisms can take up N as either nitrate or ammonium, most of the organic N within living cells is in a reduced form (ammonia). Accordingly, the mineralization of dead organic material releases NH_3/NH_4 (*ammonification*).

While the atmosphere contains half of Earth's nitrogen, the N_2 gas is rather inert and inaccessible to most organisms. To be incorporated into biomass, it needs to be "fixed" into some other form, a process that requires large amounts of energy and can be performed only by select groups of organisms. After the invention of the Haber–Bosch process in 1910, nitrogen fixing bacteria and achaea were joined in that select club by *Homo Sapiens*, a species that now accounts for over 40% of all biospheric N_2 fixation [78].

As no significant N-bearing minerals can form diagenetically within sediments, we do not need to concern ourselves with precipitation reactions. For the same reason, N transformations do not leave direct traces in sedimentary records – a complication for paleoceanographic reconstructions.

Test your understanding: In Fig. 6.2, identify the zones within the sediment where each nitrogen compound is produced or consumed. Use Fig. 6.1 to hypothesize about the likely redox processes. You may also find it helpful to consult the discussion on the shapes of vertical profiles in Section 11.2.2.

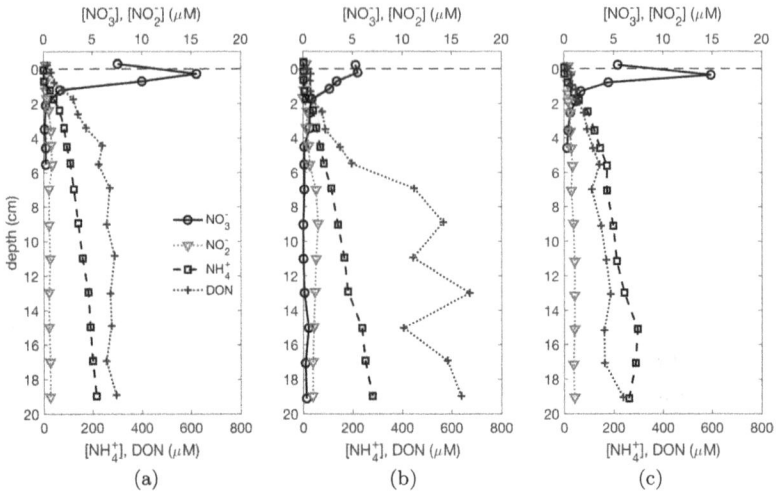

Fig. 6.2. Measured distributions of nitrogen species at several locations in Baltic Sea sediments (redrawn from [77]).

6.2 Diagenetic Reactions of N

The two most significant processes that shape the cycling of nitrogen in sediments are *nitrification* (oxidation of ammonium to nitrate), and *denitrification* (reduction of nitrate to N_2). They are major components of the global N cycle, regulating biological N availability. Other quantitatively important processes are *anammox* (anaerobic oxidation of ammonium) and *dissimilatory reduction of nitrate to ammonium* (DNRA). Fixation of N_2 is important in the water column but is rarely significant in sediments [79].

6.2.1 Mineralization of organic N

Similarly to the production of DOC and DIC through the decay of particulate organic carbon (Chapter 5), mineralization of particular organic nitrogen (PON) generates dissolved organic nitrogen (DON) and dissolved inorganic nitrogen (DIN). The latter dissolved inorganic product is mainly ammonium, which is relatively easy to measure. The buildup of ammonium in the reduced sediment porewater thus may be often used as a proxy for organic matter mineralization. The shape of the ammonium profile in the anoxic sediment often parallels the shape of the DIC profile, with the concentration ratios being similar to the typical C:N ratio in organic matter (the Redfield ratio). Unlike DIC that may precipitate as carbonate, however, ammonium is not affected by mineral precipitation. Mineralization of organic matter throughout the sediment column thus creates a distributed source of ammonium, which sustains the fluxes of ammonium upwards towards the water column and into the oxidized surface layer where it can be converted to nitrate by nitrification.

6.2.2 Nitrification

Nitrification converts ammonium into nitrate, using oxygen as the oxidant:

$$NH_4^+ + 2\,O_2 \longrightarrow NO_3^- + 2\,H^+ + H_2O$$

The process occurs near the sediment surface where oxygen is available and consumes the upward-diffusing ammonium that is produced in deeper layers. Nitrification typically proceeds in two steps, with NO_2^- as an intermediate product.

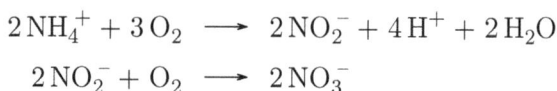

$$2\,NH_4^+ + 3\,O_2 \longrightarrow 2\,NO_2^- + 4\,H^+ + 2\,H_2O$$
$$2\,NO_2^- + O_2 \longrightarrow 2\,NO_3^-$$

The two steps are usually catalyzed by two separate groups of microorganisms. For example, ammonia-oxidizing bacteria *Nitrosomonas* may perform the first step, and nitrite-oxidizing bacteria *Nitrobacter* may use the produced nitrite to perform the second. Archaea from the *Thaumarchaeota* phylum are also known to be ammonia oxidizers. Single organisms performing a complete oxidation of ammonium to nitrate (*comammox*) have been discovered [80] but they are less efficient than dual symbiotic populations, and are consequently consigned to narrow ecological niches.[1]

---

[1]The question why the metabolic labor of nitrification is divided between two distinct populations of microbes, rather than being conducted by a single organism, probes the organizing principles of geomicrobiology. As both reactions can generate catabolic energy, one might think that a single organism would be better off performing both of them, while keeping nitrite inside the cell. Interestingly, the advantage of splitting the metabolic pathway between populations and the ecological niches for comammox were first justified theoretically [81], before the comammox bacteria were discovered experimentally.

Nitrification is typically coupled within the sediment to the opposite process, denitrification, which uses the produced nitrate.

6.2.3   Denitrification

Denitrification removes reactive nitrogen from the ecosystem by converting nitrate into biologically inert N_2. This incomplete reduction (in contrast to the complete reduction to NH_4^+) is usually coupled to the oxidation of organic carbon, thus being a form of microbial respiration:

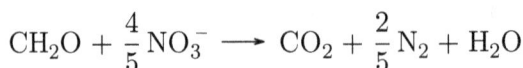

$$CH_2O + \frac{4}{5} NO_3^- \longrightarrow CO_2 + \frac{2}{5} N_2 + H_2O$$

This reaction can occur in anoxic or low-oxygen sediment layers. Denitrifying bacteria are known to tolerate low oxygen concentrations, but under oxic conditions aerobes dominate.

Denitrification occurs in multiple steps (Fig. 6.1). First, nitrate is reduced to nitrite. Then nitrite is reduced to N_2O (via NO) and then to N_2. The separate steps within the denitrification sequence may be performed by the same or different microbial populations. The nitrite generated in the first step often fuels other processes, such as anammox.

As it produces non-reactive N_2, denitrification is anaerobic the only process in the redox cascade sequence of reactions (Section 5.1.3) whose product does not consume oxygen. When the total sediment oxygen demand is used as a stoichiometric equivalent of carbon mineralization (Section 5.1.6), one should keep this in mind, allowing for an appropriate correction. As in most environments denitrification does not account for more than 5% of total C mineralization, such a correction is typically small.

6.2.4   DNRA

The dissimilatory nitrate reduction to ammonium (DNRA) is a complete reduction: it reduces the most oxidized form of nitrogen to the most reduced one:

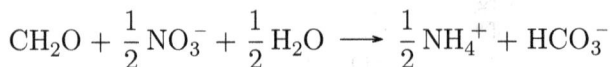

$$CH_2O + \frac{1}{2} NO_3^- + \frac{1}{2} H_2O \longrightarrow \frac{1}{2} NH_4^+ + HCO_3^-$$

Like denitrification, DNRA involves multiple steps: nitrate is first reduced to nitrite and then to ammonium (Fig. 6.1). The produced nitrite, as well as ammonium, may be used to fuel other processes, such as anammox. In contrast to denitrification, DNRA does not remove the reactive nitrogen from the system. Systems with higher prevalences of DNRA, therefore, recycle N more efficiently.

While DNRA is commonly a heterotrophic microbial metabolism coupled to the oxidation of organic carbon, it may also involve other electron donors, such as ferrous iron [82], sulfide, or elemental sulfur. The chemolithoautotrophic microorganisms that use such reactions for catabolism then use ambient CO_2 as a source of carbon.

6.2.5 Anammox

The anaerobic ammonium oxidation (anammox) oxidizes ammonium using another nitrogen species, nitrite, as an oxidant:

$$NO_2^- + NH_4^+ \longrightarrow N_2 + 2\,H_2O$$

Like denitrification, it is a pathway that removes biologically available nitrogen from the system as N_2. In the global ocean, anammox has been suggested to account for up to 50% of the removal of fixed nitrogen [83]. Significant anammox rates have also been reported in freshwater [84, 85].

As the anammox reaction does not involve organic compounds, unlike denitrification or DNRA, it does not presuppose them to be a source of carbon for anabolism. By using autotrophy (i.e. obtaining their carbon from ambient CO_2), the anammox bacteria are able to compensate for their smaller energy of catabolism by gaining competitive advantage in environments where organic carbon is scarce, such as in organic-poor pelagic sediments.

6.2.6 Adsorption of ammonium

Being a charged species, NH_4^+ may be partially absorbed by sediment minerals, as some mineral surfaces have uncompensated negative charges. Ammonium in marine sediments, for example, weakly adsorbs to clays with a linear adsorption coefficient of about 1.3 [86].

The adsorption is pH-dependent, as the pH affects the ammonia-ammonium speciation.

6.3   Nitrogen Cycling

Let us now put the above-described processes together, in the context of a redox-stratified sediment (Fig. 6.3). The organic matter, with its nitrogen content approximated by the Redfield ratio, rains down from the water column on sediments. As organic matter is mineralized by a suite of reactions described in Chapter 5, the breakdown of N-bearing compounds releases ammonium into the sediment porewater. In the sediment's oxidized zone, it is oxidized aerobically first to nitrite and then to nitrate by nitrification. Below the depth of oxygen penetration, the produced ammonium accumulates in the reduced porewater and diffuses upward.

Nitrate from the oxygenated overlying water and the uppermost sediment diffuses downward into the reduced sediment. There, it is reduced by denitrification to N_2, and to some degree by DNRA back to ammonium. Both denitrification and DNRA proceed via nitrite (NO_2^-) as an intermediate product. This nitrite may be used to oxidize ammonium in the anammox process, which converts both nitrogen compounds into N_2. The inert nitrogen gas produced by denitrification and anammox then can diffuse out of the system, returning nitrogen to the atmosphere.

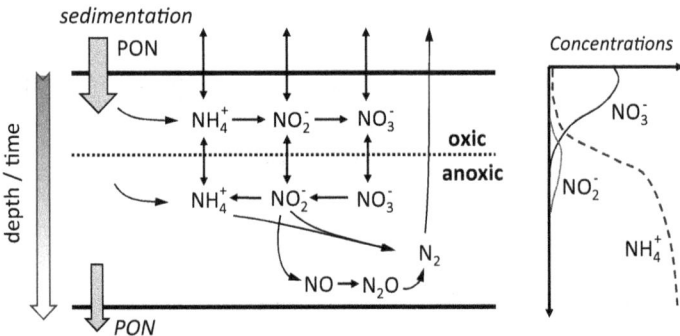

Fig. 6.3. Nitrogen cycling in sediments and the resultant typical vertical distributions of the reactive species.

Test your understanding: In Fig. 6.3, identify which transformations correspond to the individual processes discussed in Section 6.2.

Nitrification in the oxic sediment and the processes in the reduced sediment are coupled and affect each other's rates. As reactive species diffuse across the oxic-anoxic boundary, intense cycling of N takes place in the vicinity of that boundary. In particular, nitrification is fueled by ammonium that diffuses upward from the reduced sediment, and denitrification rates are in turn supported by nitrate that diffuses downward from the oxic zone.

The distributions of nitrogen species and the rates of their reactions are affected by the degree of sediment oxygenation. In organic-rich sediments where oxygen penetrates into the sediment only by a few millimeters, most of the nitrate for denitrification is supplied into the reduced sediment directly from the water column by diffusion or bioirrigation. The porewater concentrations of nitrate in such sediments decrease monotonously from the sediment-water interface, reaching zero within the denitrification zone a few mm below the depth of oxygen penetration (Fig. 6.4). In this case, the sediment serves as a sink of nitrate for the water column. In deeply oxygenated sediments, in contrast, porewater nitrate profiles are shaped by the oxidation of ammonium, which creates a peak in the nitrate profile within the oxidized sediment (Fig. 6.4). In that case, the concentration gradient across the sediment-water interface is reversed, and diffusion makes the sediment a source of nitrate to the water column. Denitrification in the reduced sediment in that case is supported entirely by the nitrate produced by nitrification, which diffuses downward from the peak [31, 87].

This difference in the provenance of nitrate also affects the way denitrification rates scale across environments (Fig. 6.5). In environments with shallow oxygen penetration, denitrification rates are generally proportional to the availability of organic carbon, and hence vary approximately linearly with the total oxygen uptake (TOU). In deeply oxygenated sediments, in contrast, organic mineralization regulates not only the supply of organic substrates for denitrification but also the supply of nitrate, through the supply of ammonium for nitrification. The denitrification rates thus become more sensitive

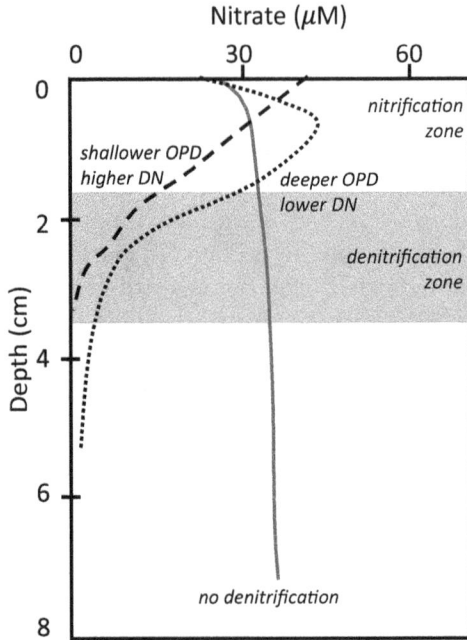

Fig. 6.4. Typical profiles of nitrogen species in sediments at sites with different organic sedimentation and/or oxygen penetration (redrawn after [31,88]).

to organic sedimentation, scaling approximately quadratically with TOU [31].

Denitrification tends to be the dominant pathway of nitrogen removal from sediments. It is particularly prevalent in coastal zones with well-oxygenated bottom water where organic matter availability and nitrate concentrations are high. The relative significance of the N removal pathways, however, varies with environmental conditions. The contribution of anammox increases (up to 20–30%) in organic-poor sediments, such as those found at greater ocean depths or in oligotrophic lakes [84], where heterotrophic processes become disadvantaged. The contribution of DNRA rarely exceeds a few percent [89], but in some environments it may account for a significant fraction of nitrate reduction [90, 91]. This may be especially true in poorly oxygenated environments where sulfur or iron species can serve as electron donors.

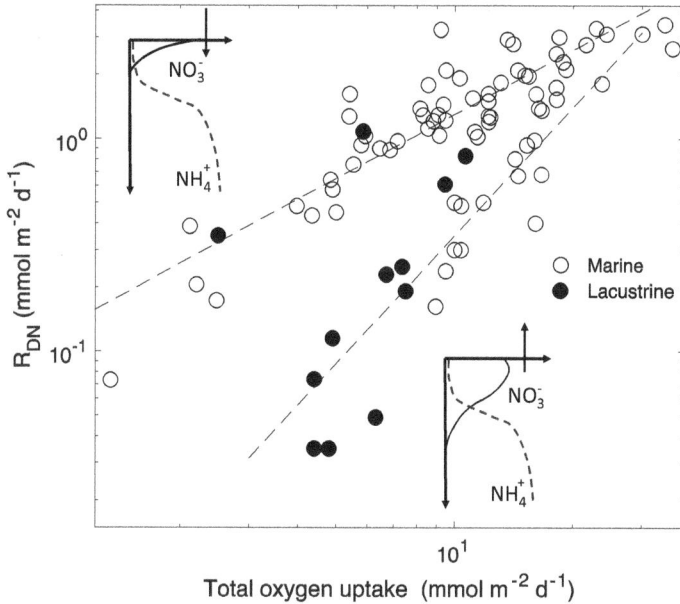

Fig. 6.5. Denitrification rates in sediments and their dependence on sediment oxygenation. Dashed lines on this logarithmic scale correspond, respectively, to linear (slope=1; at about the Redfield C:N ratio) and quadratic (slope=2) dependencies on the rates of organic carbon mineralization (redrawn from [31]).

Experimentally, characterizations of rates of the sedimentary N transformations are helped by the *isotope pairing* technique, which is often used on whole sediment cores [92]. The ^{15}N-labeled nitrate is added to the water overlying the sediment. As benthic denitrification produces N_2, where both nitrogen atoms come from the labeled nitrate, it results in $^{30}N_2$. Anammox produces N_2, where only one of the N atoms comes from nitrate, whereas another comes from the unlabeled ammonium, thereby producing $^{29}N_2$. DNRA transfers the label into the generated $^{15}NH_4$. Reoxidation of the unlabeled ammonium and the subsequent use of the produced nitrate in denitrification produces $^{28}N_2$. Measuring the abundances of each of these products in sediment incubations allows quantification of the respective process rates.

Whereas the diagenetic cycling of N is most connected to the cycling of carbon and oxygen, in some contexts its interactions with the cycles of other elements may also be important. Similarly to the iron- or sulfur-dependent DNRA, chemoautotrophic denitrification can be also coupled to the oxidation of sulfur or iron [90]. Due to such processes, the depth of nitrate penetration, rather than oxygen penetration, may mark the depth below which Fe^{2+} accumulates in sediment porewater [31] (Chapter 7). Another potentially important but so far poorly quantified connection is the role of the oxidized nitrogen species in the oxidation of methane, in a process called nitrate- or nitrite-dependent anaerobic methane oxidation (*N-damo*) [79].

6.4 Exercises

1. Synthesis exercise: Sketch the biogeochemical cycling of N for each of the following environments: (a) a well-oxygenated sediment, (b) an anoxic iron-rich sediment, and (c) an anoxic sulfur-rich sediment.
2. Which of the following processes involving nitrogen are likely to occur in sediments: N fixation, N reduction coupled to Fe oxidation, N oxidation coupled to Fe reduction, N reduction coupled to S oxidation? Explain conditions that favor or disfavor each process.
3. What is the importance of sediment denitrification for nutrient levels and primary productivity in lakes and oceans?
4. Which reaction generates more energy per nitrogen atom, nitrification or denitrification? Calculate the Gibbs free energy changes for the respective reactions under both standard and typical *in situ* conditions.
5. For the concentration gradients of nitrate and ammonium in Fig. 6.2(a), estimate approximately the sediment's rates of (a) nitrification; (b) denitrification.

Chapter 7

The Iron and Manganese Cycles

Iron and manganese are sometimes referred to as "major metals", underscoring their high natural abundances relative to other, "trace", metals, which are found at lower concentrations. The cycling of iron and manganese is strongly coupled to the cycles of other biologically important elements, particularly phosphorus and sulfur, and mediates some of the key geochemical transformations. The two metals are treated together in this chapter because their chemistry is rather similar and generates qualitatively similar patterns of cycling within the sediment. The Fe and Mn cycling is presented here in a sequence that is somewhat the reverse of the methodological order in the previous chapter. The observed vertical distributions of the solid and dissolved phases inside the sediment are presented first, with the goal for the reader to use the geochemical knowledge from the previous chapters to deduce the principal processes that may have shaped those distributions. The chemical forms of iron and manganese and their main reactions are reviewed next, followed by a reconstruction of their interactions that make up the diagenetic cycling of these metals.

7.1 Distributions of Mn and Fe in Sediments

We already saw some of the effects that sediment redox zonation exerts on the vertical distribution of redox-sensitive metals in sediments (e.g., Figs. 3.12 and 5.4). Figure 7.1 further illustrates some typical distributions of both dissolved and solid-phase iron and

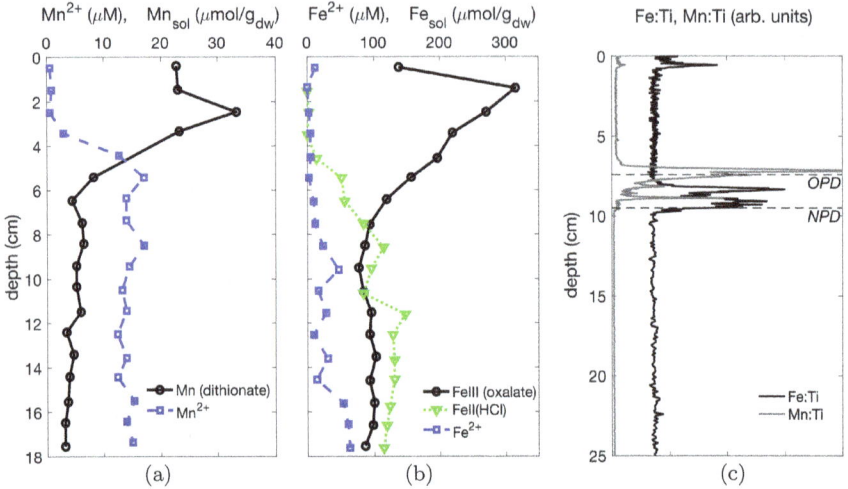

Fig. 7.1.   Distributions of Fe and Mn in sediments. (a, b) Extractable solid fractions and porewater concentrations of Mn and Fe in a Lake Michigan sediment (redrawn from [93]). (c) X-ray fluorescence (XRF) profiles of total Fe and Mn (normalized to the detrital fraction using titanium, Ti) in a Lake Superior sediment (Sta. ED, redrawn from [94]).

Note: OPD = oxygen penetration depth; NPD = nitrate penetration depth.

manganese that result from their redox interactions. Notably, the profiles of solid Fe and Mn rise sharply around the depth of the sediment redox boundary, and fall in the reduced sediment below. The concentrations of dissolved phases (Fe^{2+} and Mn^{2+}) increase in the reduced sediment, whereas their concentrations in the oxidized sediment, in contrast, are near zero.

Test your understanding: Using the profiles in Fig. 7.1 and the information in Section 3.7, which zones within the sediment serve as sources of Fe(III), Fe^{2+}, Mn(IV), Mn^{2+}, and which serve as sinks?

The previous chapters have hinted at some of the key processes that shape these distributions. The low redox potential Eh (Fig. 3.4) in deeper sediment makes reductive transformations (from Fe(III) to Fe(II), or from Mn(IV) to Mn(II)) thermodynamically favorable (Chapter 3). Their energy can fuel microbial reduction of Mn(IV) oxides and Fe(III) oxyhydroxides, coupled to oxidation of organic carbon (Chapter 5). The high redox potential in the presence of oxygen or nitrate, on the other hand, favors the reverse transformation, and can drive oxidation of the reduced metals by one of the available electron acceptors near the sediment surface. The shapes of the observed profiles, such as in Fig. 7.1, allow us to infer a number of things about the corresponding biogeochemical cycles (see also Sections 3.7 and 11.2.2):

- Dissolved Mn^{2+} and Fe^{2+} are produced in the reduced sediment, below the depth of oxygen penetration, and consumed near the boundary between the oxidized and reduced zones. In deeper sediment, Fe^{2+} concentrations are often found to be decreasing with depth, indicating a (weaker) sink for the dissolved iron there.
- Reducible solid-phase Mn(IV) and Fe(III) are consumed in the reduced sediment, as indicated by their concentrations decreasing with depth. Solid-phase Mn(IV) and Fe(III) are, however, produced around the oxic-anoxic boundary, as indicated by their enrichments there to levels that exceed those near the sediment-water interface.

The redox chemistry suggests a plausible mechanism: When solid oxides are buried below the redox boundary, their reduction creates Mn(II) and Fe(II), which are soluble and exist in aqueous solution in ionic forms Mn^{2+} and Fe^{2+}. As Mn^{2+} and Fe^{2+} become mobilized into the reduced sediment porewater, the resultant concentration gradients cause them to diffuse upward. Above the redox boundary, their reoxidation produces Mn(IV) and Fe(III) solids again (though not necessarily in their original mineral forms). These oxides subsequently can be slowly buried again into the reduced sediment, repeating the cycle. Iron and manganese thus can cycle across the oxic-anoxic boundary multiple times. The spatially separated reduction and oxidation reactions both proceed with substantial energy gain, creating multiple opportunities for microbial metabolisms. Whereas the cycling of manganese broadly mimics that of iron, the reduction and oxidation of Mn happen at a higher redox potential (Fig. 3.4). As a consequence, transformations of manganese occur higher in the sediment column than those of iron. The subsequent sections of this chapter fill in the chemical details for the cycling of the two metals.

Test your understanding: Which chemical species may potentially serve as oxidants for the reduced forms of Fe and Mn, and which can serve as reductants for the respective oxidized forms? Are there any notable differences between the redox reactions for the two metals?

7.2   Biogeochemical Reactions of Fe and Mn

7.2.1   Chemical forms of Fe and Mn

Iron is found in sediments predominantly in two redox states: ferrous (+2) and ferric (+3). Ferric (oxidized) iron is found in many common minerals, such as magnetite and hematite (Appendix B). Particularly important in sediments, however, are oxyhydroxides: goethite ($Fe(OH)_3$), ferrihydrite (α-FeOOH), and lepidocrocite (γ-FeOOH). Fresh precipitates such as ferrihydrite, rather than forming well-developed crystals, form small (nanophase) particles, which have very high specific surface area, often with unbalanced electrical charges. This makes them highly reactive and exceptionally good at adsorbing charged chemical species from the porewater (see, e.g., Chapter 8 and Section 11.3.5). At circumneutral pH, all ferric minerals are rather insoluble in water. For example, the solubility of Fe^{3+} in equilibrium with ferrihydrite is 10^{-8} M in seawater, and even lower in freshwater [95]. Ferrous iron, in contrast, is soluble, and can be present in free ionic form as Fe^{2+} at tens or hundreds of micromolar, or even higher. Ferrous minerals include siderite, vivianite, and several mineral forms of iron sulfides (Appendix B). Mixed-valence minerals, which combine Fe(II) and Fe(III), also exist. Green rust, for example, may mediate key interactions in select situations [96], though it is likely of minor importance in most modern sediments. Iron-bearing organic complexes have also been described. As they may stabilize Fe species, rendering them less reactive, they may potentially affect vertical fluxes of iron, including exchanges between sediments and the water column [97, 98].

Similarly, manganese oxides (Mn(IV) and Mn(III)) are practically insoluble, whereas the reduced Mn(II) is soluble in water and can be present as free aqueous Mn^{2+}. Manganous (Mn(II)) minerals include rhodochrosite (a manganese carbonate, $MnCO_3$), which forms at high concentrations of carbonate ions at high pH (Table C.3), and alabandite (MnS), which is highly soluble under reducing conditions [99] and therefore not commonly considered. The traditional view of the diagenetic cycling of Mn focuses primarily on the redox cycling between the Mn(IV) oxide MnO_2 and Mn^{2+}. Recently, however, the contributions of Mn(III) in both mineral and soluble forms [100] have been recognized [101, 102]. As an intermediate redox form, Mn(III) can act as either an electron acceptor or an electron donor, and can also disproportionate (transfer electrons internally, generating both reduced and oxidized products).

As catchment rocks usually contain at least an order of magnitude more Fe than Mn, concentrations of Fe in sediments are typically higher than those of Mn. At about 50-fold lower abundance in the continental crust than iron (whose weight fraction in the crust is about 5%), manganese is still the second most abundant redox-active metal in sediments, however.

7.2.2 Operationally defined fractions

Concentrations of solid phases in sediments often have to be obtained not for well-defined chemical substances but rather for the operationally defined fractions, which are extracted when the sediment is treated with certain chemical reagents [103, 104]. *Sequential extractions* treat the sediment with progressively more aggressive chemicals, so that the most reactive or easily extractable phases are extracted first, followed by the more refractory compounds. A number of extraction schemes have been developed to characterize the sedimentary pools of Fe and Mn. Most of these extractable pools contain a subset of the metal-bearing minerals. Some of the commonly considered operationally defined fractions of iron are listed in Table 7.1. Analogous operational fractions and the corresponding sequential extraction procedures for manganese may be found, for example, in [105, 106].

Test your understanding: Given the conceptual framework for the cycling of Fe outlined above, how would you expect the operational fractions listed in Table 7.1 to be distributed vertically within the sediment?

7.2.3 Fe and Mn reduction

As discussed above, bacterial reduction of Fe and Mn plays a prominent role in sediment diagenesis, being an important contributor to the oxidation of organic carbon. Mn and Fe reductions are often

Table 7.1. Selected operationally defined fractions of iron.

Fraction	Description	Ref.
0.5 M HCl-extr. Fe	"biologically available" iron: amorphous Fe(III) and solid Fe(II).	[107, 108]
Ascorbate-extr. Fe	"Loosely sorbed" Fe(II) on mineral surfaces.	–
MgCl-extr. Fe	"Exchangeable Fe": loosely sorbed Fe	[104]
Acetate-extr. Fe	"Carbonate Fe": siderite and ankerite	[104]
Dithionite-extr. Fe	"Reduceable" Fe(III) oxides: ferrihydrite, lepidocrocite, goethite, hematite, akageneite.	[104]
Fe_{HR}	"Highly reactive" iron. Classification scheme was developed for geological samples of sedimentary rocks. Fraction is assumed to include reactive Fe oxides, sulfides, and carbonates.	–
Fe_{pyr}	"Pyrite" iron; stable sulfide minerals.	–
Fe_{tot}	"Total" iron in the sample.	–

conducted by the same organisms. Manganese- and iron-reducing bacteria are metabolically versatile, capable of utilizing a range of organic substrates, such as acetate, fatty acids, lactate, formate, glucose, and others [109]. Using acetate as an example, dissimilatory reduction may be written as

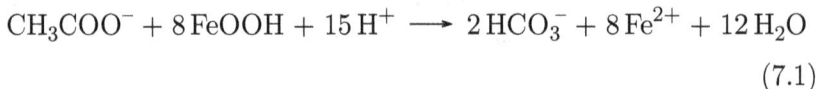

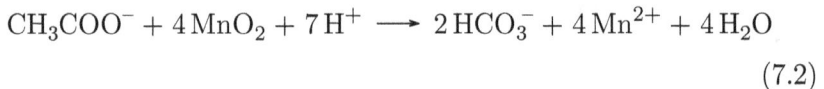

$$CH_3COO^- + 8\,FeOOH + 15\,H^+ \longrightarrow 2\,HCO_3^- + 8\,Fe^{2+} + 12\,H_2O \tag{7.1}$$

$$CH_3COO^- + 4\,MnO_2 + 7\,H^+ \longrightarrow 2\,HCO_3^- + 4\,Mn^{2+} + 4\,H_2O \tag{7.2}$$

Some organisms perform incomplete oxidation, e.g. converting lactate to acetate. Others (e.g., *Geobacter* spp.) can oxidize their substrate completely to CO_2. Besides organic substrates, growth of Fe and Mn reducers may be also supported lithotrophically, with hydrogen (H_2) as the electron donor (Table D.1). Fe and Mn reducers are also capable of utilizing alternative electron acceptors: oxygen, nitrate, sulfite, thiosulfate, and especially elemental sulfur (S^0) [6,95]. Some organisms, such as *Schewanella* spp., are facultative anaerobes, meaning they can also function in oxic environments. Such versatility allows these microbes to inhabit ecological niches in sediments outside the narrow zone of Mn and Fe reduction. Importantly for the geochemistry of trace metals, the Fe- and Mn-reducers are also capable of reducing other metals and metalloids, such as uranium(VI), chromium(VI), cobalt(III), and arsenate [6].

Accessing solid electron acceptors presents a rather special challenge for microorganisms, and typically requires direct contact of the cell with the mineral. The solubility of the mineral affects the rate of the overall reduction reaction, so the kinetics of the reduction correlates with the solubility of the solid iron phase (Table C.3). Ferrihydrite and amorphous Fe(III) oxides support the highest specific rates of reduction (highest v_{max}), whereas lepidocrocite and hematite are reduced more slowly [27,110]. The more soluble minerals additionally tend to have greater specific surface areas (m^2/g) [27]. The type of the Fe mineral being reduced also strongly affects the amount of free energy released in the reaction [111] and thus the amount of energy accessible to microbes.

Besides organic matter, iron in sediments can be reduced by several other chemical species. Hydrogen sulfide (H_2S), in particular, can reduce Fe(III) abiotically, producing elemental sulfur (S^0):

$$2\,FeOOH + 2\,H_2S \longrightarrow 2\,FeS + S^0 + 4\,H_2O \qquad (7.3)$$

Hydrogen sulfide can also reduce manganese oxides.

In addition, Fe(III) can be reduced by methane (CH_4) and, under some conditions, by ammonium (NH_4^+) (depending on the pH, the mineral phase of Fe(III) and the exact product of the ammonium oxidation; see Exercise 2 at the end of the chapter). These processes are referred to, respectively, as the *Fe-dependent anaerobic oxidation of methane (Fe-AOM)* and *Fe-ammox*:

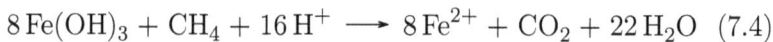

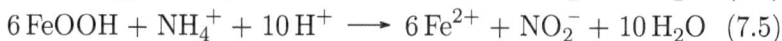

$$8\,Fe(OH)_3 + CH_4 + 16\,H^+ \longrightarrow 8\,Fe^{2+} + CO_2 + 22\,H_2O \quad (7.4)$$
$$6\,FeOOH + NH_4^+ + 10\,H^+ \longrightarrow 6\,Fe^{2+} + NO_2^- + 10\,H_2O \quad (7.5)$$

The quantitative significance of these processes in sediments is still being established [112].

7.2.4 Fe and Mn oxidation

Dissolved ferrous iron (Fe^{2+}) can be oxidized abiotically by oxygen, leading to precipitation of solid Fe(III) minerals. The abiotic reaction is fairly rapid, occurring in fully oxygenated water within minutes.

Oxidation, however, can be accelerated even further by microbes that manage to outcompete the abiotic reaction to extract energy for their metabolisms. Microaerophilic Fe(II)-oxidizing bacteria are lithoautotrophs that harness this reaction:

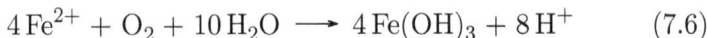

$$4\,Fe^{2+} + O_2 + 10\,H_2O \longrightarrow 4\,Fe(OH)_3 + 8\,H^+ \qquad (7.6)$$

Oxidation happens in a narrow sediment zone where Fe^{2+} is present in the same depth interval as oxygen, typically at low μM concentrations.

Nitrate, another potent oxidizer, can also oxidize Fe^{2+} (Table D.1). In sediments with wide redox zones, Fe^{2+} profiles can be seen extending from the reduced sediment up to the penetration depth of nitrate, rather than that of oxygen [31]. Nitrate may be reduced either to N_2 or NH_4^+, similarly to its reduction in denitrification vs. DNRA (Chapter 6). The reduction may be chemolithoautotrophic, using ambient CO_2 as a carbon source for microbial growth. More commonly, however, nitrate is used to oxidize organic carbon, while Fe^{2+} is oxidized by the produced nitrogen species, such as nitrite (in a so called "chemodenitrification") [109]. Oxidation of iron with nitrate or nitrite is a relatively slow process, however.

Manganese oxides (Mn(IV)) also can oxidize Fe(II):

$$2\,Fe^{2+} + MnO_2 + 2\,H_2O \longrightarrow 2\,FeOOH + Mn^{2+} + 2\,H^+ \qquad (7.7)$$

The reaction is abiotic, with surface-controlled kinetics [109]. As this reaction precipitates iron oxides while mobilizing manganese oxides, it serves to minimize the overlap between the Fe-rich and Mn-rich layers within the sediment, which are often seen as distinct laminations abutting each other (Fig. 7.1).

The aerobic oxidation of Mn(II)

$$2\,Mn^{2+} + O_2 + 2\,H_2O \longrightarrow 2\,MnO_2 + 4\,H^+ \qquad (7.8)$$

yields less energy than the aerobic oxidation of Fe(II), and proceeds at slower rates (below pH 8). Incomplete oxidation of Mn(II) can result in Mn(III), which may either precipitate (e.g., as MnOOH) or disproportionate into Mn^{2+} and MnO_2. Although most Mn oxidation in sediments is microbially-catalyzed, manganese oxidizers are organoheterotrophs, apparently not gaining energy from the reaction [6].

7.2.5 Mineral precipitation

Diagenetic formation of minerals in the reduced sediment is a pathway by which iron can be buried and removed from circulation. Reduced Fe-bearing minerals include iron sulfides, iron carbonates (siderite), and iron phosphates (vivianite).

In marine sediments, sulfides are the main sink for iron. Dissolved Fe^{2+} readily precipitates in the presence of dissolved hydrogen sulfide, forming solid iron monosulfides:

$$Fe^{2+} + HS^- \longrightarrow FeS + H^+ \tag{7.9}$$

Subsequent diagenetic reactions typically transform FeS into the more stable mineral pyrite (FeS_2). The reaction is often written as involving either elemental sulfur or additional hydrogen sulfide:

$$FeS + S^0 \longrightarrow FeS_2 \tag{7.10}$$
$$FeS + H_2S \longrightarrow FeS_2 + H_2 \tag{7.11}$$

Given the need for an oxidant to transition from FeS to FeS_2, the last reaction could, perhaps, be written in combination with hydrogenotrophic methanogenesis as

$$4\,FeS + 4\,H_2S + CO_2 \longrightarrow 4\,FeS_2 + CH_4 + 2\,H_2O \tag{7.12}$$

Formation of siderite ($FeCO_3$)

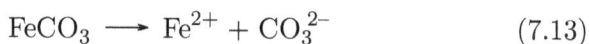

$$FeCO_3 \longrightarrow Fe^{2+} + CO_3^{2-} \tag{7.13}$$

occurs at fairly high carbonate ion concentrations (see Table C.3 and Exercise 3 at the end of the chapter), and so is more readily achieved in sediment porewater than in the overlying water column [113]. Fe may also become incorporated into other carbonate minerals, such as calcite, as a trace element. The iron phosphate mineral vivianite ($Fe_3(PO_4)_2$) is an important sink for sediment phosphorus (Chapter 8). It is common in lake sediments and was also found to be abundant in some marine sediments [114].

Concentration profiles of dissolved manganese, Mn^{2+}, often flatten out in the deep sediment, indicating the absence of sinks. In some marine sediments, however, the buildup of manganese can lead to supersaturation with respect to manganese carbonates [69].

Manganese carbonates—rhodochrosite ($MnCO_3$) and kutnohorite ($Mn_xCa_{1-x}CO_3$)—may then form (Table C.3). A particularly interesting form of Mn-rich mineral formation is the growth of manganese (or ferromanganese) nodules at the surface of deep sea sediments. Nodules are centimeter-scale spheroidal or discoidal concretions that form on the surface of marine sediments, and occasionally in freshwater lakes [115,116]. Growing by mechanisms that are still not fully resolved, they consist of oxidized crystalline and amorphous forms of Mn (vernadite, birnessite, todorkite) and Fe oxyhydroxides [69].

7.3   Cycling of Fe and Mn in Sediments

We can now interpret the processes that shape the profiles of Fe and Mn in sediments (Fig. 7.2). Iron and manganese are deposited from the water column into the sediment as oxidized mineral particles. When buried into the reduced sediment, these oxides are reductively dissolved, primarily by microbial reduction coupled to the oxidation of organic matter. Reduction of iron oxides by H_2S generated during sulfate reduction may be important as well. The produced dissolved Fe^{2+} and Mn^{2+} then diffuse upward towards the oxic-anoxic boundary where they precipitate again as oxyhydroxides or oxides. Dissolved Fe^{2+} in deeper, reduced sediment is removed by precipitation with hydrogen sulfide. Produced FeS further matures into pyrite, which is the dominant form of sulfide preserved in the longterm. Fe^{2+} may also be incorporated in other ferrous minerals.

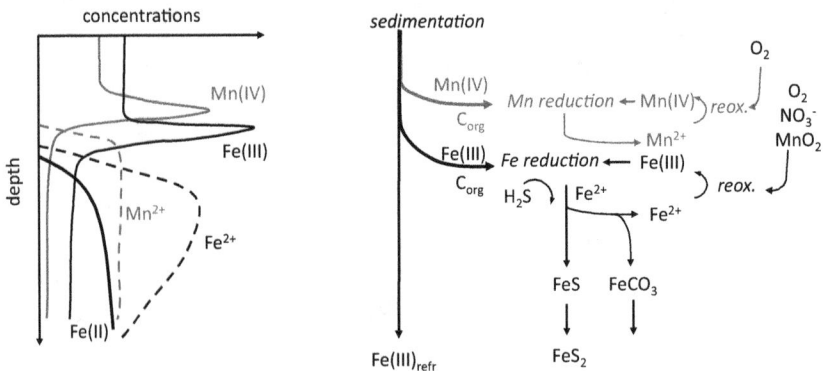

Fig. 7.2. Diagram of the diagenetic cycling of Fe and Mn.

Manganese is more efficiently recycled within sediments, with only a small proportion being buried. As Fe oxides are less soluble than Mn oxides, and aqueous Fe(II) is oxidized faster than Mn(II), diagenetic redistribution of Mn is more extensive than that of Fe [95], and a higher proportion of iron is retained in sediments than Mn.

The higher redox potential for Mn reduction than for Fe reduction (Fig. 3.4) leads to a vertical separation of the corresponding diagenetic zones. Together with the possibility of Fe(II) being oxidized by Mn(IV), this leads to a vertical separation of the corresponding diagenetic enrichments, with Mn-rich layers forming above Fe-rich layers.

As Fe and Mn can cycle around the redox boundary repeatedly, their atoms are effectively recycled, so the diagenetic reactions that they support can be repeated many times. A single atom of Fe or Mn therefore may be used to oxidize dozens of carbon atoms in organic matter, despite the nominal stoichiometries (4Fe:1C and 2Mn:1C) suggested by reactions 7.1–7.2.

7.4 Exercises

1. Synthesis exercise: Compare the cycling of Fe in sediments when the bottom water contains oxygen vs. anoxic conditions. Draw a diagram illustrating the cycling identify the differences in the dominant chemical forms of iron and their vertical fluxes within the sediment and across the sediment-water interface.
2. Which Fe-ammox reactions are energetically favorable at pH 7? Which mineral phases of Fe(III) are they more likely to involve and what are the possible oxidation products of ammonium?
3. What are the typical concentrations of Fe^{2+} and DIC for which sediment porewaters would be oversaturated with respect to siderite?
4. In the event when the concentration of oxygen in the overlying water decreases, so the oxic-anoxic boundary moves closer to the sediment-water interface, which metal-rich layer – Mn or Fe – would dissolve quicker? Why? What would happen to the mobilized fractions? Would such an event leave any geochemical signatures in the sediment that could be preserved for several years or decades?

Chapter 8

The Phosphorus Cycle

Phosphorus (P) is a key nutrient for aquatic organisms. In many environments, its availability to photosynthetic organisms determines the rates of carbon fixation and thus defines the biological productivity of the entire ecosystem. This is especially true in freshwater, where overabundance of P can lead to poor water quality and algal blooms. The biological productivity in surface waters, in turn, determines the amount of organic debris that settles into the deeper waters, and thus the rates of oxygen consumption there and the flux of particulate organic matter to the bottom.

Sediments serve as important sinks for phosphorus. They can, however, become sources of P, particularly during episodes of anoxia when phosphorus is mobilized from them as dissolved orthophosphate. The efficiencies of phosphorus mobilization are determined by biogeochemical reactions within the sediment, and the permanent burial of P into the deep sediment depends on the formation rates and stability of P-bearing solid phases. By regulating P supply to the water column, these processes impact the entire ecosystem. This chapter reviews the chemical forms of phosphorus in sediments and the processes that move phosphorus chemically between its various forms and physically within the sediment. It also discusses factors that control the availability of the sedimentary P to the overlying water column.

8.1 Phosphorus in Sediments

Unlike nitrogen, phosphorus has no gas phase. It comes into aquatic bodies from the watershed with organic matter or minerals, carried by runoff or dust. While organisms in the water column may recycle it multiple times, phosphorus is only removed from the aquatic body with the outflow or into the sediments. The sedimentary sink often dominates (Fig. 8.1) and thus controls the mass balance and the total amount of P circulating through the water column.

Phosphorus is essential to all life because of its role in key molecules, including DNA, RNA, ATP, and phospholipids. In waters where P availability throttles the biological productivity of phytoplankton, organisms can draw the concentration of bioavailable P (orthophosphate) well below 1 μM, under the detection limit of typical analytical techniques [118, 119]. Where P is scarce, such as in oligotrophic lakes, its scarcity restricts biological productivity and increases water transparency. Over-enrichment in P leads to eutrophication and excessive primary production, especially of algae and cyanobacteria [120]. In the ocean, productivity is more strongly regulated by the availability of nitrogen, with estuaries and the continental shelf sometimes exhibiting mixed limitation by N and P [120].

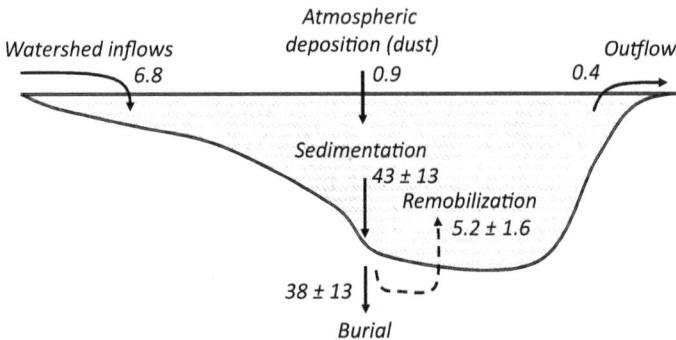

Fig. 8.1. Significance of sediments as sinks of phosphorus in a large oligotrophic lake (Lake Superior). All fluxes are normalized to μmol/m^2/day [117]. In dynamic systems with long histories of P deposition and long time scales of sediment diagenesis, instantaneous budgets of incoming and outgoing fluxes may be substantially unbalanced.

8.1.1 Chemical forms of P

Unlike other diagenetically important elements, phosphorus exists in sediments only in one oxidation state, P(V), and almost exclusively as one type of chemical functional group: orthophosphate $(-PO_4^{3-})$. Accordingly, it does not undergo redox transformations. Free orthophosphate dissociates in water into four aqueous species: H_3PO_4, $H_2PO_4^-$, HPO_4^{2-}, and PO_4^{3-}. Only $H_2PO_4^-$ and HPO_4^{2-}, however, are present in significant concentrations at environmentally relevant pH values (Fig. C.4). This dissolved phosphate is the only form of P that can be assimilated by phytoplankton [120], and is therefore highly bioavailable.

Organic matter incorporates phosphate groups as components of cellular material. Fresh marine phytoplankton is usually thought to be characterized by a relatively constant ratio of C:P [121] (the Redfield ratio, about 106:1 mol/mol), though the reality appears to be more complicated (Table 10.1). The phytoplankton C:P ratio varies more strongly in freshwater. There, the Redfield ratio is rather an exception on the low end of the spectrum than the rule [122, 123]. The ratio may also change seasonally in response to changes in P limitation [124]. Zooplankton have higher phosphorus content relative to carbon [123] than phytoplankton. While their C:P ratios in freshwater typically still exceed the canonical Redfield ratio, they vary in a wide range (120–230), being strongly affected by variations in nutrient demand. Some, like *Daphnia*, may have ratios as low as 50 [124] (Table D.1). Organisms still higher up on the food chain, like fish, have even higher C:P proportions in their tissues.

Common P-bearing minerals (Appendix B) include apatite (calcium phosphate) and its variations, especially fluorapatite, which forms diagenetically in marine sediments. Another diagenetically important phase is vivianite $(Fe_3(PO_4)_2)$, a ferrous iron phosphate [125, 126]. Phosphorus may also be incorporated as a trace element into other minerals, such as calcite [127].

Test your understanding: If no phosphate could diffuse out of the sediment or otherwise escape from the porewater into which it is mobilized by mineralization of organic matter, what would be the maximum concentration of the dissolved phosphate? Assume a sediment that contains 1 wt% of organic carbon, composition of the organic matter being that of marine phytoplankton; porosity of 0.9, solid sediment density of 2.65 g/cm^3, and the mineralization efficiency of both organic C and organic P at 50%. (The answer is on page 194.)

8.1.2 Operationally defined fractions

The concentrations of phosphorus in the environment are typically measured in operationally defined fractions. Dissolved phosphate that is readily available to photoplankton is usually characterized as *soluble reactive phosphorus*, SRP. It is assumed to consist of dissolved orthophosphate (which forms phosphomolybdic acid in a traditional colorimetric "molybdenum blue" method of measurement). Total dissolved phosphorus (TDP) is operationally defined as phosphorus that passes through a fine filter (typically 0.45 μm). Besides dissolved ionic phosphate, TDP samples may also include colloidal fractions and certain organic molecules. Total phosphorus (TP) is the total amount of phosphorus in the sample, in both dissolved and particulate forms, usually after larger organisms such as zooplankton are filtered out.

Similarly to the operationally defined fractions for solid-phase iron (Table 7.1), phosphorus in the sediment's solid fraction is also characterized in fractions that can be extracted with different reagents, usually as a sequential extraction. A popular SEDEX extraction scheme [128, 129], for example, distinguishes between loosely sorbed P, ferric iron-bound P, authigenic carbonate-associated P, detrital apatite, and organic P (Table 8.1). Phosphorus also can be measured

Table 8.1. Selected operationally defined fractions of phosphorus.

Fraction	Description	Ref.
SRP	"Soluble reactive phosphorus" – dissolved orthophosphate and colloidal fractions that pass through filter	—
MgCl$_2$-extr.	"Exchangeable" or "Loosely sorbed" P.	[129]
CDB-extr.	"Fe-bound P" – Phosphorus bound with reducible iron.	[129]
Acetate-extr.	"Carbonate P". Authigenic apatite (CFA), biogenic apatite, CaCO$_3$-bound P.	[129]
1M HCl-extr.	Detrital apatite, other inorganic P	[129]
550°C, 1M HCl	"Organic P"	[129]
TP	"Total phosphorus" – all P in the sample	

in some of the same fractions as Fe [94], which helps identify the degree to which the two elements are linked.

The standard method for analyzing phosphate in water samples and extracts is the molybdenum blue method [130]. It allows detecting phosphate concentrations down to about 0.2 μM. More complicated methods, such as the more sensitive magnesium-induced coprecipitation (MAGIC) technique [119, 131], can detect concentrations of phosphate down to tens of nM.

8.1.3 The association with iron

Of primary importance for the diagenetic cycling of P is the ability of dissolved phosphate to become immobilized on the surfaces of iron oxyhydroxides (Fig. 8.2). Freshly precipitated amorphous Fe(III) minerals, in particular, have a high specific surface area (m^2/g) [27] and a high density of positively charged surface sites. Negatively charged phosphate molecules can easily adsorb to them. Phosphorus can also become trapped within crystals. In deeper, reduced

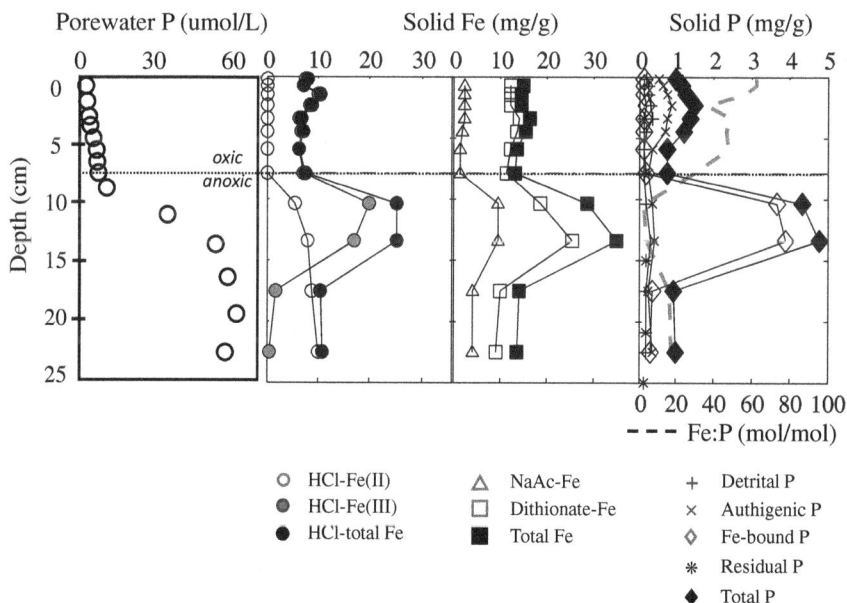

Fig. 8.2. Depth distributions of phosphorus-bearing fractions in a Lake Superior sediment (adapted from [117]).

sediment, a major Fe-bearing mineral containing P is vivianite, particularly in freshwater and brackish environments [114].

The cycling of P thus becomes coupled to the cycling of Fe. Whereas P itself does not undergo reduction or oxidation, the reductive dissolution of iron minerals in anoxic sediment can release phosphate into solution. Conversely, oxidation of Fe^{2+}, followed by precipitation of fresh Fe(III) oxyhydroxides near the sediment's oxic-anoxic boundary, can bind phosphate into the solid phase. Estimates for the Fe:P ratios in such precipitates vary, with sorption capacity reaching its limit at Fe:P ratios around 7–12 (mol/mol) [117]. Sediments of oligotrophic lakes that are iron-rich may have an order of magnitude higher ratios ([117, 132]). In sediments with low inventories of reactive Fe(III), mineral surfaces become saturated and hence cannot efficiently bind P and prevent phosphorus from diffusing past the oxic-anoxic interface into the overlying water [133–135].

8.2   Phosphorus Cycling

Being a major component of organic matter, phosphorus becomes deposited into sediments with organic debris (Fig. 8.3). The mineralization of organic matter hydrolyzes phosphate compounds and releases them into the porewater. Their subsequent fate depends on the geochemical cycling of multiple elements. While phosphate itself does not participate in many reactions, its association with iron recycles it across the sediment's redox boundary. The cycling of iron, in turn, is tightly coupled to the cycles of oxygen, nitrate, and sulfur (Fig. 8.3). The net result – the fraction of phosphate that reaches the overlying water or becomes permanently buried into deep sediment – is determined by the balance among several processes: the release of dissolved phosphate from organic matter, immobilization of Phosphate with Fe near the sediment surface, immobilization of P with reduced minerals in the deeper sediment, and the upward diffusion of phosphate [136].

Adsorption of phosphate onto Fe(III) oxyhydroxides often leads to P being enriched around the oxic-anoxic boundary, at the depth that matches the diagenetic Fe enrichments (Fig. 8.2). Phosphate diffusing from deeper, reduced, sediment layers becomes trapped there, increasing the P:Fe ratio (Fig. 8.2). Only a fraction of phosphate penetrates the oxidized layer, which limits P effluxes into the overlying water. Phosphate may traverse the oxygenated layer more easily when the surfaces of Fe-bearing minerals become saturated and can no longer adsorb additional phosphate.

As burial transports the Fe(III) solids into the reduced sediment, their reductive dissolution re-mobilizes phosphate there, increasing its porewater concentrations (Fig. 8.2). This stimulates the upward diffusion of phosphate towards the oxic layer, perpetuating the cycle. The situation mirrors the diagenetic cycling of Fe, which is recycled multiple times across the redox boundary. The mobilized P can cycle between the oxidized and reduced sediment zones until it is either immobilized within solid minerals in the reduced sediment or escapes into the overlying water column (Fig. 8.4).

Several processes can remove Fe^{2+} from circulation, making it unavailable for reoxidation and the adsorption of P. Hydrogen sulfide (H_2S), in particular, reacts with Fe(III) phases to reductively dissolve them, and precipitates with the produced Fe^{2+} to form insoluble iron sulfides (Fig. 8.3) (Chapter 7). This decreases the efficiency of P immobilization.

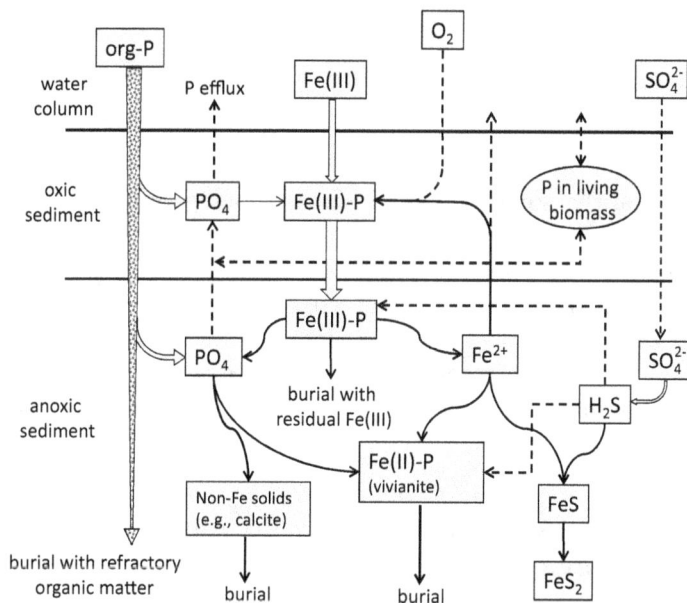

Fig. 8.3. Phosphorus cycling in sediments (adapted from [137] with permission from Taylor and Francis Group LLC).

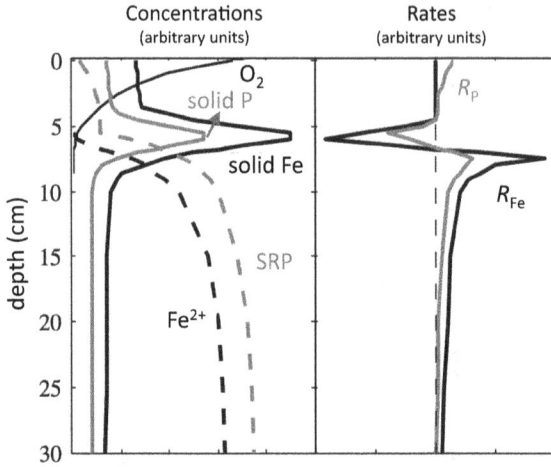

Fig. 8.4. General shapes of the vertical distributions of P- and Fe-bearing phases, and the associated depth variation in the volume-specific rates of production (positive) or consumption (negative) of the porewater phosphate (R_P) and dissolved ferrous iron (R_{Fe}). Positive R_P values near the sediment-water interface are associated with the release of phosphate during mineralization of freshly deposited organic matter (adapted from [117]).

Test your understanding: How would the efficiency of P immobilization be affected by the concentration of sulfate (SO_4^{2-}) in the overlying water?

Importantly, some of the phosphate, being bioavailable, may be incorporated into the living biomass of benthic microbes. Some of them accumulate polyphosphates as energy reserves [138]. This biological P can be actively exchanged with both porewater and overlying water [139, 140]. In some cases, benthic macrofauna may also become a significant reservoir of sedimentary P, as is the case with the invasive quagga mussels in the lower four of the North American Great Lakes [141].

8.2.1 Releases of P from sediments during seasonal anoxia

Phosphorus can become mobilized from sediments into the overlying water during periods of anoxia, when Fe oxides become reductively dissolved. This is a frequent, if not regular, occurrence in many productive temperate lakes where summer stratification of the water column prevents oxygen from reaching the bottom. In the 1930s, Wilhelm Einsele [142] and Clifford Mortimer [143] observed that P was being released from lake sediments into the overlying water shortly after the onset of anoxia (Fig. 8.5). This mobilization of phosphorus occured immediately following the release of ferrous Fe^{2+} from sediments. Such observations led to what became known as the "classical paradigm" of P releases from lake sediments. The paradigm, which stated, in essence, that phosphorus effluxes were linked to oxygen concentrations, dominated limnological thinking for many decades, creating a firm (and often misleading) association between phosphate mobilization and anoxia. It underpinned, for example, attempts to prevent P releases from sediments by artificially oxygenating the deep waters of lakes, which were not always successful [144].

A better understanding of the mechanisms by which P mobilization occurs [136, 146] helped clarify some important nuances of this phenomenon [137]. While maintaining an oxidized surface layer provides a mechanism for intercepting the upward fluxes of P from the deeper sediment, such a layer, to be efficient, requires adequate

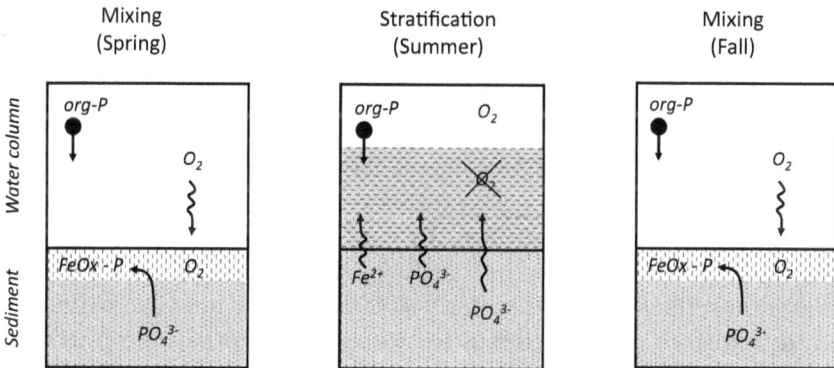

Fig. 8.5. The "classical paradigm" of P releases from sediments during seasonal anoxia in lakes. In temperate climates, thermal stratification in the water column during the summer prevents oxygen from reaching the deep waters, causing anoxia. Reductive dissolution of iron oxides in the sediment then removes the barrier to the P upward diffusion and mobilizes the previously accumulated phosphorus. The classical paradigm, however, may not apply on longer time scales or in systems where the iron oxides near the the sediment-water interface are already saturated with phosphate [137, 145].

amounts of reactive Fe(III). P fluxes may be unaffected if Fe(III) surfaces become saturated with phosphate. In addition, regulation of P fluxes happen differently on seasonal vs. multi-year time scales [136, 145]. While short-term anoxia may release P from sediments by dissolving Fe(III) oxides, the reverse process takes place when the bottom waters are reoxygenated at the end of the season. Long-term P retention requires a permanent removal of phosphorus into the deep sediment, either with organic matter or in mineral phases. Even when seasonal P releases are prevented, long-term fluxes may be unaffected if P is not buried into the deep sediment. In short, what is not buried will eventually be released (Section 11.2.1). The Fe(III) layer will eventually become saturated and P will leak out into the water column. Oxygenating the hypolimnion thus can only work as a short-term solution [133, 136]. Long-term solutions to curtailing phosphorus concentrations in lakes thus involve binding phosphorus with non-redox sensitive substances, such as alum, which maintain their adsorptive processes under the reducing conditions in the deep sediment. The most efficient long-term solution to combat P-induced eutrophication is, however, to restrict the inputs of phosphorus from the watershed, even in situations with high internal (i.e., from sediments) fluxes. On the multi-year time scales when the mass balance of P in the lake can be simplified by neglecting seasonal dynamics, the concentrations of total phosphorus in the water column may be predicted from the external inputs and the efficiency of P recycling in sediments [145].

In deeply oxygenated sediments, the redox cycling of Fe may be occurring too deep below the interface to affect the exchanges of P across the sediment surface. In such situations, P fluxes across the interface are governed primarily by the mineralization of organic matter and the associated hydrolysis of phosphate near the sediment-water interface. The P effluxes then become essentially proportional to the rates of organic carbon mineralization [117].

As P fluxes from sediments can strongly influence the biological productivity in the water column, which in turn can influence the efficiency of P retention in sediments, there is a potential for positive feedback between the P cycling in the water column and its cycling in sediments [136]. High biological productivity leads to bottom anoxia, poor P retention, and consequently even more nutrients for primary producers. In contrast, low productivity may lead to better burial of

P with Fe oxides (though less burial of P with organic matter, which is more completely mineralized under oxygenated conditions). Such positive feedbacks may lead to hysteresis or even bistability in the ecosystem's dynamics.

8.3 Exercises

1. Synthesis exercise: Sketch and compare diagrams for the diagenetic cycling of P under persistently anoxic conditions, persistently oxic conditions with deep oxygen penetration, persistently oxic conditions with shallow oxygen penetration, and periodically anoxic conditions.
2. Synthesis exercise: Analyze how the efficiency of P immobilization in sediments may be affected by (a) the sulfate concentration; (b) the organic carbon concentration; (c) the oxygen concentration.
3. What are the typical concentrations of phosphate (in its most abundant ionic forms) and dissolved iron that would favor the formation of vivianite? How would the favorability of vivianite formation vary with sediment pH?
4. Some sediment contains a 0.5 cm-thick diagenetically formed Fe(III)-rich layer near its oxic-anoxic boundary. The sediment porosity is 0.9, the solid sediment density is $\rho = 2.5$ g/cm^3, and the average concentration of Fe within the layer is 1 wt%. The Fe(III) phases within the layer are at their saturation limit for adsorbing dissolved P. Estimate the average fluxes of phosphate $(\mathrm{mmol\,m^{-2}\,d^{-1}})$ from this sediment into the overlying water during an episode of a prolonged anoxia, during which the entire Fe-rich layer reductively dissolves over a period of 1 month.
5. *Bonus question*: For problem 4 above, is the 1-month time scale for the dissolution of the Fe-rich layer realistic? Consider typical processes by which such dissolution may occur, typical rates of such reactions, and the typical composition of the sediment with respect to the relevant chemical species.

Chapter 9

The Sulfur Cycle

Sulfur (S) is abundant in the environment in both aqueous and mineral forms. Ranging in oxidation states from -2 (sulfide) to $+6$ (sulfate), it is an essential component of all living cells, and a substrate for numerous catabolic reactions. Sulfur links multiple other biogeochemical cycles, affects mineralization of organic matter, mediates the cycling of iron and phosphorus, controls fluxes of methane, and affects the activities of benthic animals and nearly any other aspect of diagenesis. In past geologic epochs, marine sulfur cycling regulated the temperature and the atmospheric oxygen concentration on the planet [147]. It caused mass extinction events, and left in the rock record some of the most useful paleo-markers that we use today to reconstruct ancient marine conditions. In seawater, sulfate is the second most abundant anion, after chloride. The foul-smelling hydrogen sulfide produced by bacteria from sulfate is highly reactive, driving numerous oxidation reactions. It is also toxic to eucaryotes, affecting habitability of benthic environments. Sulfur-reducing bacteria mediate important environmental processes, including methylation of mercury. Mobilization of sulfur from deposited organic debris may sustain diagenetic reactions, whereas the post-deposition sulfurization of organic matter may preserve it against degradation, leading to formation of stable hydrocarbons such as kerogens.

Despite this multitude of connections, the cycling of sulfur in aquatic sediments is often presented in a rather simple fashion, while a deeper analysis can reveal a dizzying amount of detail. In describing key reactions and processes of the sulfur cycle, this chapter first presents a simplified version, followed by a gradual buildup of complexity.

9.1 A Simplified Sulfur Cycle

The transformations of two aqueous species, sulfate (SO_4^{2-}) and hydrogen sulfide (H_2S/HS^-), and of the solid iron sulfides, FeS and FeS_2, have already been mentioned in the preceding chapters. Together, they capture many of the key features of the diagenetic S cycling, including its interactions with organic matter (Chapter 5) and iron (Chapter 7). Some of the most salient features of S cycling can be obtained by placing the already mentioned reactions within the context of a redox-stratified sediment column, as illustrated in Fig. 9.1.

In this simplified version of the sulfur cycle, the key biogeochemical transformations are as follows.

- Dissimilatory reduction of sulfate to hydrogen sulfide during anaerobic mineralization of organic matter in the reduced sediment:

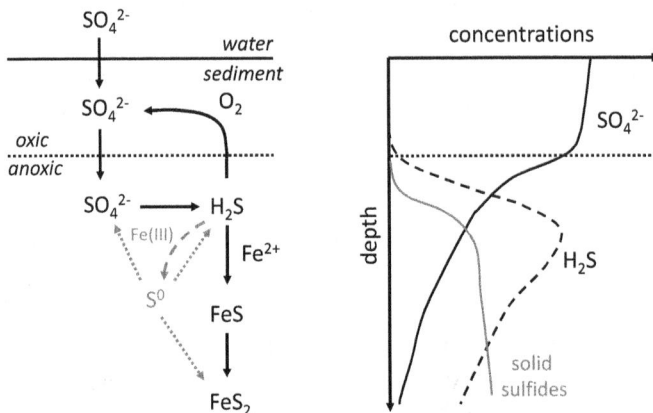

$$CH_2O + \frac{1}{2}SO_4^{2-} \longrightarrow HCO_3^- + \frac{1}{2}H_2S \qquad (9.1)$$

Fig. 9.1. A simplified sulfur cycle: dissimilatory sulfate reduction, sulfide reoxidation, and precipitation of solid sulfides. The simplified reaction network may also include disproportionation of elemental sulfur and the sulfate-dependent anaerobic oxidation of methane (AOM).

The reaction is carried out by a diverse community of sulfate-reducing bacteria. The sulfate is sourced from the overlying water column via diffusion across the topmost oxidized layer. The hydrogen sulfide is produced by sulfate reduction in a sediment zone that begins below the penetration depths of oxygen and nitrate, and typically below the narrow zones of manganese and iron reduction.

- Reoxidation of sulfide back to sulfate by oxygen in the oxic surface layer:

$$H_2S + 2O_2 + 2H_2O \longrightarrow SO_4^{2-} + 2H^+ \tag{9.2}$$

This oxidation reaction recycles back to sulfate the hydrogen sulfide that diffuses upwards across the redox boundary. The reaction is highly energetic, so sulfide is reoxidized with high efficiency, limiting the fluxes of sulfide into the upper strata. Oxidation of sulfide may be similarly coupled to the reduction of nitrate, another potent oxidant (see below).

- Immobilization of sulfide in solid minerals, typically by co-precipitation with Fe^{2+} as monosulfides or pyrite:

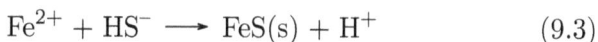

$$Fe^{2+} + HS^- \longrightarrow FeS(s) + H^+ \tag{9.3}$$

The reaction serves as a long-term sink for the sedimentary sulfur, leading to its burial. It limits the concentration of dissolved hydrogen sulfide, which, in the absence of dissolved iron, may accumulate in the reduced sediment porewaters. In marine sediments, where sulfate reduction greatly outpaces iron reduction, sulfide accumulates in the sediment below the iron reduction zone, where Fe^{2+} concentrations become low. In sulfate-poor freshwater environments, Fe^{2+} often dominates, and sulfide is present at low concentrations only near the locus of its production within the sulfate reduction zone.

As sulfate enters the sediment from the overlying water and is consumed in the reduced sediment, its concentration decreases with depth below the sediment-water interface. In marine environments where the sulfate concentration is high (28 mM in seawater), the decrease is relatively slow, so the vertical profiles of sulfate extend deep into the sediment. In freshwater sediments, sulfate is often exhausted several centimeters below the oxygen penetration depth. The concentration profiles of hydrogen sulfide often exhibit peaks around the sulfate reduction zone, as sulfide is reoxidized above that zone and consumed by mineral precipitation (or other processes) below it.

Several additional redox transformations become important in certain contexts. Reduction of sulfate coupled to the oxidation of methane (the anaerobic oxidation of methane, AOM [28]) consumes sulfate at the lower end of its concentration profile (which coincides with the upper end of the methane profile):

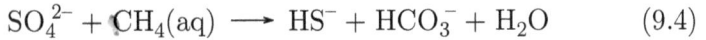

$$SO_4^{2-} + CH_4(aq) \longrightarrow HS^- + HCO_3^- + H_2O \qquad (9.4)$$

In marine sediments, the reaction may occur several meters below the sediment surface but nevertheless serves as an important barrier to the diffusion of methane from deeper sediments. In freshwater sediments, at low sulfate concentrations, the reaction is only marginally feasible thermodynamically, and is almost never reported.

At depths where oxygen and nitrate are unavailable, dissolved hydrogen sulfide may be quickly and efficiently oxidized (abiotically) by reactive iron oxides:

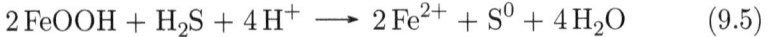

$$2\,FeOOH + H_2S + 4\,H^+ \longrightarrow 2\,Fe^{2+} + S^0 + 4\,H_2O \qquad (9.5)$$

The reaction is fast and may compete with the microbial reduction of iron oxides. The produced Fe^{2+} becomes available for precipitation of solid sulfides, further increasing the demand on porewater H_2S. Rather than generating sulfate, the reaction produces elemental sulfur (S^0), which can undergo further transformations. In particular, S^0 can disproportionate to produce both sulfate and sulfide (see below), or become incorporated into pyrite.

The simplified representation shown in Fig. 9.1 is sometimes sufficient for analyzing vertical distributions of the most abundant sulfur species and for understanding the rates of the associated biogeochemical transformations. The sulfur cycle, however, can be significantly more complex, as detailed below.

9.2 Biogeochemical Reactions of S

9.2.1 Chemical forms of S

Sulfate ($+6$) and hydrogen sulfide (-2) are, respectively, the most oxidized and the most reduced endmembers in a rather long set of

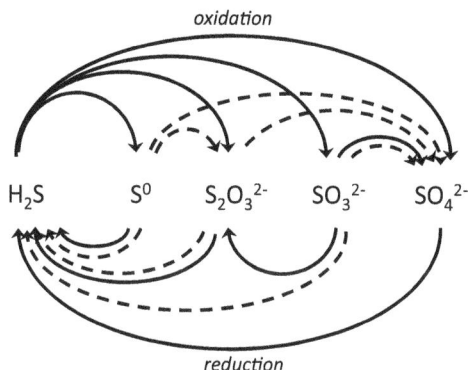

Fig. 9.2. Sulfur species and redox transformations. Dashed lines indicate disproportionation.

reactive sulfur species (Fig. 9.2). Inorganic sulfur species of intermediate valences include sulfite (SO_3^{2-}, +4), thiosulfate ($S_2O_2^{2-}$, +2), and elemental sulfur (S^0, 0). Elemental sulfur, being a common intermediate product of sulfide oxidation, can be as abundant as solid iron sulfides (Fig. 9.3), reaching in marine sediments $mmol\,L^{-1}$ levels. Dissolved sulfite and thiosulfate exist at porewater concentrations that rarely exceed several μM. They are highly reactive, being able to participate in both oxidation and reduction reactions, as well as disproportionation (Fig. 9.2), and are quickly consumed. The diagenetic sulfur cycling also includes other reactive species, such as trithionate ($S_3O_6^{2-}$, +3.33) and tetrathionate ($S_4O_6^{2-}$, +2.5), which are also capable of participating in metabolic reactions [148].

9.2.2 S reduction and oxidation

As sulfur atoms travel up and down the redox ladder of their multiple oxidation states, their reduction and oxidation reactions form a dense and interconnected network (Fig. 9.2). Participation of intermediate species allows for complex reaction pathways [149,150]. A large proportion of hydrogen sulfide produced by bacterial sulfate reduction is reoxidized to several compounds in intermediate redox states. The produced intermediates are, in turn, highly reactive and undergo further transformations. The energetic gains for some of the reactions are illustrated in Table 9.1. The high energies of reactions, particularly for thiosulfate, cause the intermediate species to be in high

Table 9.1. Selected reactions of sulfur oxidation, reduction, and disproportionation. The Gibbs free energies (kJ/mol) are calculated per S-atom and per electron transferred, at pH 7, $T = 4°C$, and for the following ambient concentrations: 10 μM for acetate, thiosulfate, hydrogen sulfide, Fe^{2+}, and Mn^{2+}; 300 μM for sulfate; 3 μM for sulfite; 20 μM for nitrate; 0.1 μM for H_2. Goethite was used as the reactive Fe(III) phase.

Reaction	$\Delta G/S$	$\Delta G/e^-$
Reduction with acetate		
$CH_3COO^- + SO_4^{2-} \longrightarrow HS^- + 2\,HCO_3^-$	−31	−3.9
$CH_3COO^- + 2\,SO_4^{2-} + H^+ \longrightarrow S_2O_3^{2-} + H_2O + 2\,HCO_3^-$	+4.1	+1.0
$CH_3COO^- + S_2O_3^{2-} + H_2O \longrightarrow 2\,HS^- + 2\,HCO_3^- + H^+$	−35	−8.9
$CH_3COO^- + 4\,S^0 + 4\,H_2O \longrightarrow 4\,H_2S + 2\,HCO_3^- + H^+$	−23	−11
Reduction with H_2		
$SO_4^{2-} + 4\,H_2 + H^+ \longrightarrow HS^- + 4\,H_2O$	−167	−21
$S_2O_3^{2-} + 4\,H_2 \longrightarrow 2\,HS^- + 3\,H_2O$	−14	−3.4
$S^0 + H_2 \longrightarrow H_2S$	−4.8	−2.4
Oxidation with nitrate		
$\frac{8}{5}NO_3^- + H_2S \longrightarrow \frac{4}{5}N_2 + \frac{4}{5}H_2O + SO_4^{2-} + \frac{2}{5}H^+$	−696	−87
$NO_3^- + H_2S + H_2O \longrightarrow SO_4^{2-} + NH_4^+$	−471	−59
Oxidation with Fe(III) and Mn(IV)		
$MnO_2 + H_2S + 2\,H^+ \longrightarrow Mn^{2+} + S^0 + 2\,H_2O$ *(inorg.)*	−135	−68
$2\,FeOOH + 2\,H_2S \longrightarrow S^0 + 2\,FeS + 4\,H_2O$ *(inorg.)*	−57	−29
$2\,FeOOH + H_2S + 4\,H^+ \longrightarrow 2\,Fe^{2+} + S^0 + 4\,H_2O$ *(inorg.)*	−58	−29
$8\,FeOOH + S_2O_3^{2-} + 14\,H^+ \longrightarrow 2\,SO_4^{2-} + 8\,Fe^{2+} + 11\,H_2O$	−184	−46
$8\,FeOOH + 2\,HS^- + 16\,H^+ \longrightarrow S_2O_3^{2-} + 8\,Fe^{2+} + 13\,H_2O$	−173	−43
Disproportionation		
$4\,S^0 + 4\,H_2O \longrightarrow SO_4^{2-} + 3\,HS^- + 5\,H^+$	−16	−8.0
$4\,S^0 + 3\,H_2O \longrightarrow S_2O_3^{2-} + 2\,H_2S + 2\,H^+$	−7.1	−3.5
$S_2O_3^{2-} + H_2O \longrightarrow SO_4^{2-} + H_2S$	−40	−9.9
$4\,SO_3^{2-} + 2\,H^+ \longrightarrow 3\,SO_4^{2-} + H_2S$	−209	−35

biological demand. Thiosulfate, in particular, being an important product of sulfide oxidation and itself a potent oxidant as well as a reductant, may "shunt" the S cycle [151, 152], serving as an intermediate node for the reaction network. It is highly bioavailable and has been shown to be active in both marine and freshwater environments, supporting bacterial populations even in sulfur-poor sediments.

The most abundant reductant of sulfur is typically organic matter, in the form of low-molecular-weight compounds, such as acetate or lactate. Hydrogen (H_2) is also extensively used by both heterotrophic and autotrophic microbial populations. The primary source for the reduced and intermediate sulfur species into the sediment is the reduction of the most abundant oxidized form of sulfur, sulfate, which is used as an electron acceptor to oxidize organic matter. While most of the sulfate reduction is performed by microorganisms as a purely catabolic reaction (dissimilatory reduction), without incorporating sulfur into the biomass, organisms can also perform assimilatory reduction, where the reduced sulfur is incorporated into the cell components. Assimilatory reduction typically involves an energy cost to the organism [148].

Oxidation of sulfur species can be linked to a number of oxidants. Oxygen, nitrate, and metal oxides can all serve as electron acceptors (Table 9.1). Sulfide oxidation with nitrate, for instance, is a highly energetic reaction exploited by a number of microbes, including giants such as the *Thioploca* spp. bacteria. Below the depth of nitrate penetration, reactive metal oxides serve as an important sink for sulfide. Dissolved hydrogen sulfide is oxidized by reactive Fe and Mn oxides quickly and abiotically. In marine environments, the half-life for the oxidation reaction may be as short as tens of minutes in the case of Fe oxides, and less than a minute for Mn oxides [153]. Oxidation of sulfide by metal oxides produces mostly elemental sulfur. Oxidation by oxygen or nitrate, in contrast, proceeds to sulfate, though thiosulfate and sulfite can also be produced as reactive intermediates.

9.2.3 S disproportionation

Disproportionation is a redox reaction in which electrons are transferred between atoms of the same element, resulting simultaneously in oxidized and reduced products. For example, elemental sulfur (e.g., as polysulfides) can disproportionate by transferring 6 electrons from one S atom and distributing them among three other S atoms, to produce one sulfate ($+6$) and three sulfide (-2) molecules (Table 9.1). Other products, including thiosulfate, are also possible. Thiosulfate and sulfite are similarly capable of disproportionation (Table 9.1). The disproportionation of sulfite is a particularly energetic reaction, even at low concentrations (Table 9.1). Disproportionation of elemental sulfur is unfavorable under standard conditions but is favorable under typical *in situ* conditions, where dissolved sulfide is scavenged [154]. Like other redox reactions, disproportionation reactions can be harnessed by microbes to power their metabolisms. Some organisms perform disproportionation as their sole metabolism, whereas many perform the reaction while also growing on sulfate reduction [148].

An opposite reaction, *comproportionation*, whereby a reduced and an oxidized species react to form a compound of the intermediate oxidation state may also be energetically feasible under certain conditions [155]. For example, H_2S and sulfate (SO_4^{2-}) may be combined to produce elemental sulfur (S^0).

9.2.4 S mineral precipitation

Sulfur can precipitate in either oxidized or reduced forms, but oxidized sulfur minerals (sulfates) rarely form sediments. Barite ($BaSO_4$) and gypsum ($CaSO_4 \cdot 2\,H_2O$), formed typically by evaporation of seawater, are usually brought into the sediments from the watershed. The most abundant reduced sulfur minerals are sulfides, particularly iron sulfides [156]. Monosulfides (nominally FeS), with the oxidation state of sulfur of -2, are metastable and can become oxidized if exposed to oxygen. The more stable pyrite (FeS_2), with the average oxidation state of sulfur of -1, is stable over the geologic time scales. Diagenetically formed elemental sulfur (S^0) is also found in sediments, sometimes in substantial quantities.

Iron sulfides form by co-precipitation of dissolved hydrogen sulfide with ferrous iron (Eq. (9.3)), which readily produces metastable iron monosulfides. Monosulfides then transform into pyrite via one of the possible pathways:

$$FeS + H_2S \longrightarrow FeS_2 + H_2 \qquad (9.6)$$

$$FeS + S^0 \longrightarrow FeS_2 \qquad (9.7)$$

The transformation effectively involves an oxidation of sulfur (from -2 to -1), thus needing either a more reduced product (H_2), or an oxidant in a higher redox state (S^0).[1]

---

[1]This reaction commonly involves polysulfides S_8 or S_x^{2-} [156, 157].

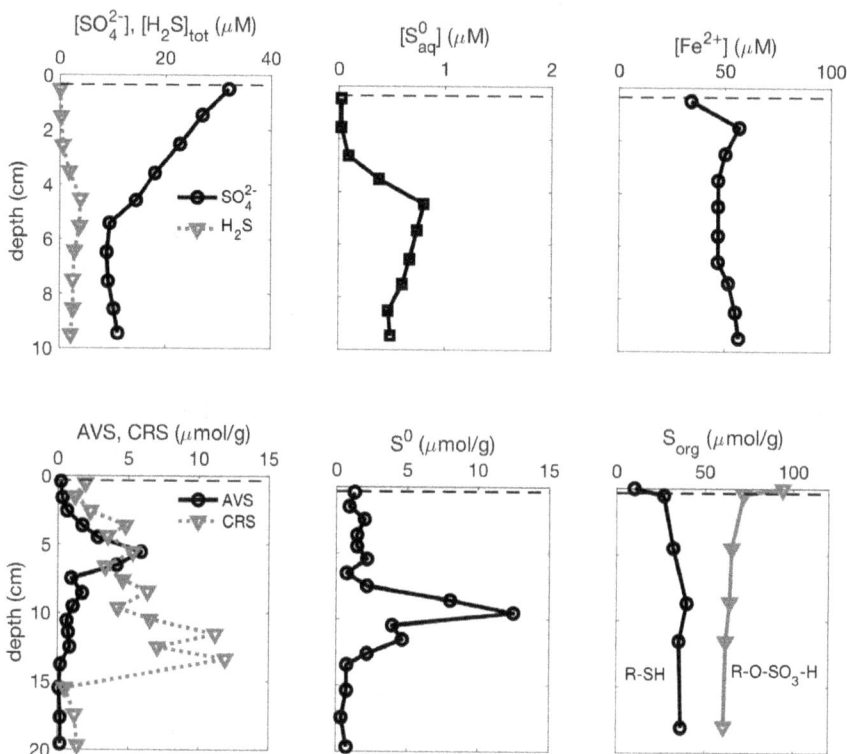

Fig. 9.3. Vertical distributions of sulfur species in sediment of an oligotrophic lake (Lake Tantaré, Quebec; redrawn from [159]). Dashed horizontal line indicates the oxygen penetration depth.

Experimentally, solid-phase sulfides are characterized in operationally defined fractions. Monosulfides are commonly associated with acid-volatile sulfides (AVS) [158]. Pyrite is included in the chromium-reducible sulfur (CRS) fraction (Fig. 9.3).

9.2.5 Organic S mineralization and post-depositional sulfurization

Sulfur makes up about 1% of fresh organic matter by weight, about the same proportion as phosphorus. It is an essential component of aminoacids (cysteine and methionine), sulfolipids, sulfate esters (choline sulfate), and cofactors (coenzymes, vitamins, electron couriers) [148]. Just as phosphate is mobilized during organic matter mineralization (Chapter 8), hydrolysis and mineralization of the organic material can mobilize substantial amounts of sulfur compounds. In marine sediments where sulfate is readily available from seawater, their amounts are fairly small relative to the amount of sulfate being drawn from the overlying water. In low-sulfate freshwater environments, however, the release of oxidized sulfur compounds and the reoxidation of the mobilized reduced S-bearing molecules can sustain vigorous sulfur cycling within the sediment, including high rates of sulfate reduction [32, 160]. In some sediments, the concentration profiles of porewater sulfate may even exhibit peaks below the sediment-water interface, indicating that sulfate is being produced within the sediment and diffuses out into the overlying water column [32]. In such cases, sedimentary sulfate reduction, rather than being fueled by sulfate from overlying water, may be fueled entirely by the internal pool of organic S [32, 161].

In deeper sediment, organic matter undergoes the process of post-depositional sulfurization. Hydrogen sulfide reacts with organic molecules, inserting sulfur atoms into the organic matter [162, 163]. Binding of sulfur with the organic matter can be a relatively fast process and, even at low concentrations of sulfide, may outpace the precipitation of iron sulfides as a diagenetic sink for sulfur [164]. Sulfurization generally results in organic matter becoming more refractory and increases its burial efficiency (Fig. 5.7).

9.2.6   Cryptic cycling

Some of the most important chemical species—those that are in high biological demand and cycle rapidly—may be present in sediment porewater at concentrations below detection limits, precisely because any time a molecule is produced it is immediately taken up. This leads to the concept of *cryptic cycling*. Several sulfur species may cycle cryptically. For example, sulfide produced through sulfate reduction can immediately react with ferrous iron, leading to non-detectable concentrations of the dissolved sulfide in the porewater. Highly reactive species, such as thiosulfate, are also likely to cycle this way. Their contributions can be revealed, however, through experiments that measure reaction rates, such as those utilizing radioactive tracers (see below).

9.2.7 Rates of sulfate reduction

Dissimilatory sulfate reduction is an anaerobic metabolism, even though some sulfate reduction has been detected in oxic environments, perhaps carried out by specially adapted microorganisms [148] or in anoxic micro-environments. Accordingly, the main zone of sulfate reduction begins in the anoxic sediment below the penetration depths of oxygen and nitrate, and typically below the zones of Mn reduction and Fe reduction. Sulfate reduction continues, at diminishing rates, into the deeper sediment until sulfate is exhausted. In marine sediments, sulfate profiles terminate at the sulfate-methane transition zone (SMTZ) several meters below the sediment-water interface [28]. When sulfate is abundant, the rates of bacterial sulfate reduction are controlled by the rates at which organic reductants are supplied by the organic matter mineralization (Section 5.2.2, Eq. (11.20)). In lake sediments, at depths where sulfate is abundant enough not to limit the reduction rates, sulfate reduction rates may be as high as in coastal marine sediments (Fig. 9.4). The shallower depth of sulfate penetration maintains a steeper gradient of sulfate, sustaining its high downward fluxes. In well oxygenated sediments, as proportionally more of the organic matter is mineralized aerobically, the contribution from sulfate reduction decreases (Fig. 9.4). As sulfate becomes depleted with depth, in freshwater organic-rich sediments its contribution is typically smaller than in marine sediments [93, 133], being replaced at depth by methanogenesis. Whereas in freshwater sediments the contribution of sulfate reduction to organic carbon mineralization rarely exceeds 20%, in coastal marine sediments contributions of 50% and above are common [134, 165].

At low sulfate concentrations, the rates of microbial sulfate reduction become limited by the availability of sulfate and the microbial abilities to access it. The typical concentrations at which the scarcity of sulfate begins to limit microbial activities (characterized by the Monod K_m parameter, Eq. (11.30)) have been variably reported at around 200 μM (mostly in marine environments) and 5 μM (mostly in freshwater). These two distinct ranges seem to correspond to two different mechanisms of microbial rate regulation

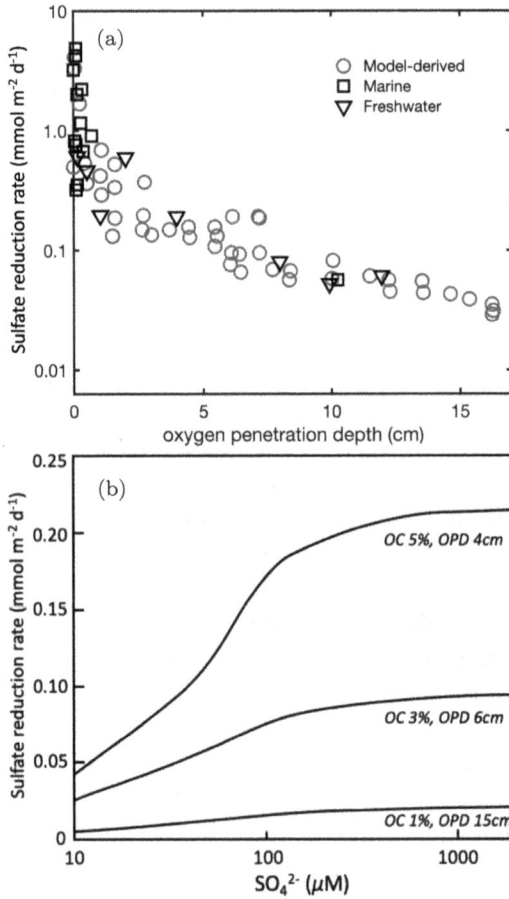

Fig. 9.4. Sulfate reduction rates in sediments, as related to (a): the depth of oxygen penetration (OPD); (b): the concentration of sulfate in overlying water and the sediment organic carbon (OC) content (adapted from [32]).

([166, 167]). Below these characteristic values, sulfate concentrations are often reported to persist into the sediment porewater at low μM levels [74]. In sedimentary environments where sulfate concentrations in the overlying water are below 100 μM, rates of bacterial sulfate reduction fall below the maximum rates of organic matter mineralization and instead depend on the concentrations of sulfate (Fig. 9.4), as sulfate reduction becomes limited by the supply rates of sulfate and microbial capabilities. At these low levels, rather than being drawn from overlying water, a substantial proportion of sulfate may come from the mineralization of organic sulfur [32].

Rates of sulfate reduction are commonly measured by injecting sulfate labeled with a short-lived radioactive isotope of sulfur, $^{35}SO_4^{2-}$, into the sediment, incubating the sediment, and determining the radioactivity of the ^{35}S in the pools of sulfate and sulfide [168]. To obtain the rates as functions of depth within the sediment, the label may be injected into intact sediment cores at multiple depth intervals.

9.2.8 Microbes involved in sulfur cycling

The dominant majority of sulfate reducers belong to the domain of *Bacteria*. They perform dissimilatory sulfate reduction using as electron donors either organic compounds or hydrogen. Large organic molecules are inaccessible to anaerobic microbes without prior fermentation, but sulfate reducers can still use a range of organic substrates, including hydrocarbons, aliphatic acids (formate, acetate, propionate, etc.), sugars, and others. While some bacteria are capable of oxidizing organic substrates, especially acetate, completely to CO_2, many perform incomplete oxidation to simpler organic molecules [148]. Many sulfate reducers also reduce elemental sulfur, and account for most of the reduction of thiosulfate and sulfite [148]. Reduction of sulfate by methane (anaerobic oxidation of methane, AOM) is carried out by a syntrophic consortium of archaea and bacteria [28].

The bacterial reduction of sulfate is usually accompanied by large isotopic fractionations, whereby sulfate molecules containing the lighter isotopes of sulfur (^{32}S) are reduced preferentially, whereas the heavier isotopes (^{33}S and ^{34}S) are left behind in the pool of unreacted sulfate (Appendix E). As a consequence, the $\delta^{34}S$ of the produced sulfide (the expressed fractionation) can be lower than that of the source sulfate by as much as 70‰. Other S reduction processes generate characteristically smaller fractionations. The isotopic signatures of S transformations become preserved in precipitated sulfide minerals and can be used to reconstruct the diagenetic microbial processes from the sediment record [169].

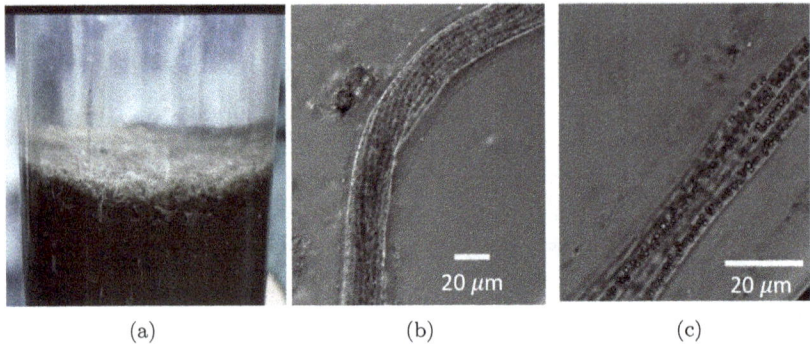

(a)   (b)   (c)

Fig. 9.5. (a) Giant *Thioploca* bacteria in a marine sediment microbial mat in the Peru–Chile oxygen minimum zone (image by Lisa Levin; NOAA). (b, c) *Thioploca* in a freshwater sediment in Lake Superior (reprinted from [170] with permission from Elsevier). Under magnification, multiple braided trichomes are visible within a sheath (b), with visible S^0 globules within them (c).

Sulfur oxidizers thrive in sulfidic sediments. Some of the most dramatic examples are the microbial mats composed of giant, cm-long *Thioploca* found in the oxygen minimum zone off the coasts of Chile and Peru (Fig. 9.5a). These bacteria use nitrate as the electron acceptor and can concentrate nitrate in a central vacuole [171]. Despite the low concentrations of sulfide in freshwater, abundant populations of *Thioploca* have also been found in lakes [170, 172] (Fig. 9.5b,c). Their metabolisms have been suggested to include the use of thiosulfate [172], while nitrate has been suggested to be reduced either to ammonium (as in DNRA [173]) or to nitrogen gas (as in denitrification [174]). Other large sulfur-oxidizers include filamentous mat-forming *Beggiatoa* spp. and *Thiomargarita*, whose spherical cells reach nearly mm size and may be visible to a naked eye. Sulfide oxidizers can typically also oxidize elemental sulfur, thiosulfate, and tetrathionate.

9.3   Exercises

1. Synthesis exercise: Sketch a diagram for the simplified sulfur cycling under (a) persistently anoxic conditions; (b) in an environment where anoxic sediment is periodically in contact with oxygenated bottom waters.

2. Synthesis exercise: Based on the concentration profiles in Figure 9.3 and assuming steady state, is it possible to estimate the total (depth-integrated) rate of sulfate reduction in the reduced sediment? Would this require any additional information about the sediment? What might be the role of sulfide reoxidation?

3. What physical, chemical, and biological factors are likely to control the rate of sulfate reduction in vertically stratified sediments? How do the relative roles of these factors depend on the type of the sedimentary environment, the concentrations of relevant chemical species, and/or the depth within the sediment? Does the reaction rate ever switch between being reaction-controlled and transport-controlled?

4. In what types of environments may sulfur comproportionation become thermodynamically favorable?

Chapter 10

Sediments as Environmental Archives

The constantly accumulating piles of sediments at the bottoms of lakes and oceans create historical archives for the markers of environmental conditions that become preserved in the sediment's solid matrix. Techniques for interpreting these records underlie the disciplines of paleoceanography and paleolimnology. Past climates are inferred from the sediments of oceans and deep ancient lakes, and records of human history are often obtained from water bodies proximate to human settlements. This chapter reviews the methods for establishing the timeline of sediment deposition and the inorganic and organic substances that can be used in the reconstructions of past histories. A sediment is not a passive repository, however: depositional signatures become altered during diagenesis. The second part of this chapter analyzes some of these changes and the methods for deconstructing them.

10.1   Sediment Dating

Interpreting sedimentary records usually requires associating specific sediment layers with the dates when they were deposited. Non-bioturbated sediments may contain annual laminations (varves), which can be counted back in time from the sediment-water interface, much like tree rings. In some sediments, specific depth intervals may be associated with well-known external events. Layers of fine volcanic particles (tephra), for example, may mark the years of eruptions, which are often known from other sources. Bulk lead (Pb) concentrations may reflect atmospheric emissions from leaded gasoline, which in the US peaked around 1975. In European lakes, sediment horizons containing radioactive cesium ^{137}Cs or strontium ^{90}Sr mark 1986—the year of the Chernobyl nuclear incident. A more diffuse cesium peak worldwide serves as a legacy of the above-ground nuclear testing that ended in 1963. More commonly, however, dating of the sedimentary archives requires the use of radioactive tracers, such as ^{210}Pb or ^{14}C.

10.1.1 ^{210}Pb dating

The radioactive isotope of lead, ^{210}Pb, is derived from the radioactive decay of radon gas (^{222}Rn), which is constantly emitted into the atmosphere from soils at a more-or-less constant rate. The fallout of ^{210}Pb from the atmosphere varies geographically because of differences in rainfall. The radioactive lead eventually makes its way into the sediments [175,176]. Under conditions of constant sedimentation, the concentration of this radioactive isotope decreases with time after its deposition with a half-life of 22.3 years (see Appendix E). Due to this relatively short decay period, ^{210}Pb can be used for dating relatively recent sediments, deposited over the past 100–150 years. In older sediments, the low concentrations of radioactive atoms result in greater uncertainties.

Because the amount of the radioactive isotope decreases with time after deposition exponentially (Appendix E), a constant burial velocity would result in an exponentially decreasing profile of the ^{210}Pb activity (decays per second) when the activity is plotted against depth within the sediment. On a log-linear plot, it would appear as a straight line (Fig. 10.1). As burial velocity usually decreases with depth because of sediment compaction (Chapter 2), a better variable for plotting, instead of depth, is cumulative mass accumulation (g/cm^2). When the flux of ^{210}Pb to the sediment surface is relatively constant over time, the ages of sediment layers can be calculated from the log-linear graph of the remaining ^{210}Pb radioactivity (Fig. 10.1). The activity decreases by a factor of two in every depth interval whose width corresponds to the half-life of the isotope. The radioactivity that is due to the deposited Pb (*unsupported activity*) is typically measured as the activity in excess of the background radiation (*supported activity*). The latter reflects the *in situ* decay

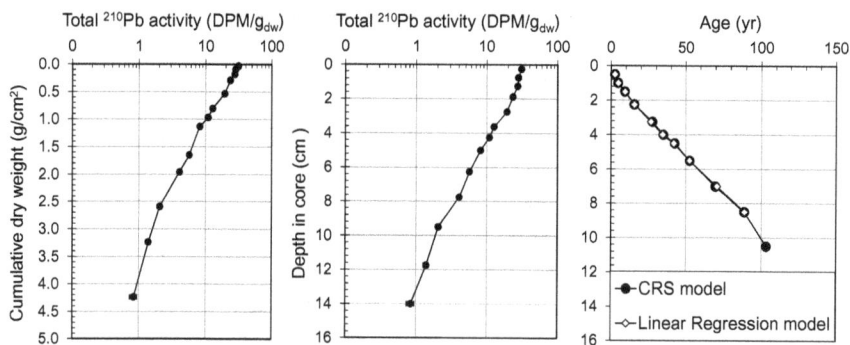

Fig. 10.1. Total ^{210}Pb activity in the sediment from the hypoxic zone (65 m depth) of meromictic Lake Matano (Indonesia), and the associated dating scheme.

of natural ^{226}Ra, and is estimated from the radioactivity in deeper sediment. As bioturbation smears the uppermost sediment layer, the exponential decay from which the sedimentation rate is inferred must be measured below the bioturbation zone.

To account for time variations in either the supply of the fallout lead or the sedimentation rate, two models are commonly used. The (more popular) constant-rate-of-supply (CRS) model assumes a constant rate of supply of fallout ^{210}Pb, irrespective of variations in sedimentation. The constant-initial-concentration (CIC) model, in contrast, assumes the supply of ^{210}Pb to vary proportionally with sediment mass accumulation [176].

Test your understanding: When the plot of the logarithm of the unsupported ^{210}Pb activity against depth within the sediment looks like a straight line, what does it tell you about the diagenetic processes and the depositional environment? Which factors could cause the plot to deviate from a straight line? What might be the effects of bioturbation or variations in porosity?

10.1.2 Radiocarbon (^{14}C) dating

Radioactive decay of ^{14}C atoms, which have the half-life of about 5730 years, enables dating of sediments thousands of years into the past. Twigs, small fragments of bark, leaves, or pine needles washed into the sediments are likely to be no more than a few years older than the sediment layer in which they became trapped.

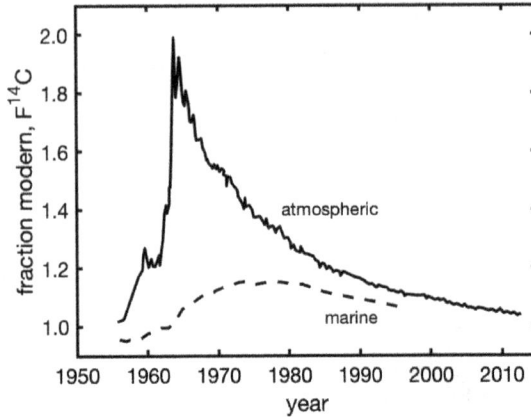

Fig. 10.2. Concentration of radioactive ^{14}C in the Earth's atmosphere and the effect of "bomb carbon" (redrawn from [177]). The fraction Modern (F^{14}C) expresses the activity of ^{14}C relative to that of the time before the nuclear tests, which is taken to be 1.

Performing radiocarbon analysis on them helps to constrain the age of the associated sediment layer. Continuous depth profiles of ^{14}C measured in the sedimentary organic matter are also routine, but interpretations need to take into account the potentially long time that the organic carbon has spent in the system before its deposition into the sediment (the *reservoir effect*).

In recent sediments, an important complication is the need to correct for the increases in atmospheric ^{14}C that occured from the mid-1950s until 1963, as a result of above-ground testing of nuclear weapons (the so-called "bomb carbon") (Fig. 10.2). Since the second half of the 20th century, the concentrations of ^{14}C in both the atmosphere and seawater have been higher than they would have been naturally. The non-monotonic increase in ^{14}C creates an ambiguity: sediments deposited in the 1960s may have the same or greater radioactivity as sediments that were laid down in later years.

10.2 Inorganic and Organic Markers of Environmental Processes

Sedimentary records are created when particles settle to form sediments and become buried. The particles come from three main sources: land (terrigenous or clastic material), the remains or metabolic products of aquatic organisms (biogenous or biogenic material, also termed "oozes"), and chemical reactions within the water body (hydrogenous material). They may include minerals, pollen, fish, remains of aquatic phytoplankton and zooplankton, aerosols, aquatic pollutants, and other chemicals [178].

10.2.1 Sources of sediment

Terrigenous inputs produce particles of varying sizes, forming sand, silt, or clay. Coarser particles are deposited shorter distances away from the shore or river mouth, whereas pelagic sediments tend to be composed of finer clays (Fig. 10.3). Alternations in sediment grain types within the same sediment may thus reflect changes in the hydrodynamic processes that transported the sediment or changes in water levels, which were accompanied by advancing or retreating shorelines. Variations in grain sizes thus provide information

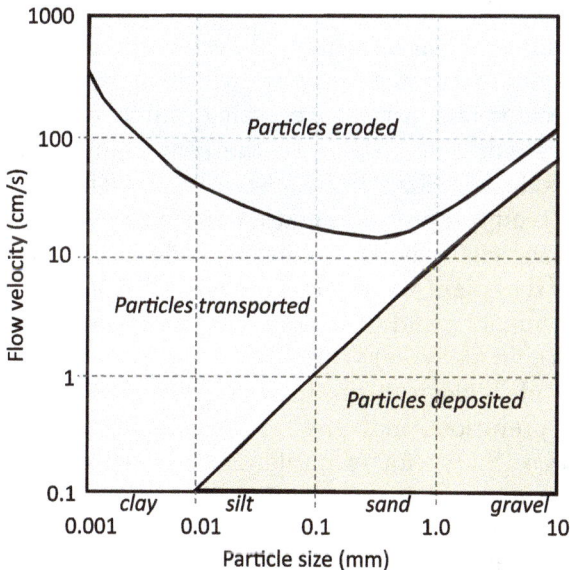

Fig. 10.3. The Hjulström curve [179] of particle erosion and deposition. Hydrodynamic flows determine the grain size of particles that can accumulate in sediments. Cohesiveness of clays increases the velocity needed for erosion.

about the rates of catchment weathering and the velocities of currents within the water body.

Biogenic particles include organic matter and the "hard parts" of organisms, such as shells, bones, or exoskeletons. Hydrogenous sediments are formed when minerals crystalize and precipitate. Their amounts may reflect changes in the chemistry or temperature of water. For example, "whiting events" in hardwater lakes happen when enhanced biological productivity during a hot summer increases the pH and causes precipitation of calcium carbonate [180].

10.2.2 Biomarkers

Biomarkers (short for biological markers) are biologically produced compounds that preserve information about the organisms that produced them. Having structures that are specific enough to be reliably associated with their originating organism, they may be used as proxies for the organisms themselves or for the processes that those organisms mediated [181]. Biomarkers may include organismal remains (e.g., fossil diatoms), pollen, hard-to-degrade molecular compounds, or other indicators (Fig. 10.4). Lipids, in particular, are slow to degrade and can be reliably detected and identified by methods such as liquid chromatography, which makes them a popular class of molecular biomarkers. Analyzing sediments for the presence of biomarkers can help determine the sources of organic material (with alkanes, fatty acids, alcohols, sterols, pigments, or intact polar lipids), past temperatures of aquatic environments (with alkenones, GDGTs, or long-chain diols), conditions and vegetation in the watershed (with $\delta^2 H$ of plant waxes, or charcoal), stratification and salinity (with tetrahymanol), and histories of anthropogenic pollution (with pollutant-specific analyses) [181].

Histories of biological productivity can also be investigated, though interpretations may need to account for multiple factors. *Biogenic silica* (SiO_2), for example, is derived from siliceous algae such as diatoms, and its profiles within the sediment have been used as indicators of overall productivity. The relationship, however, is complicated, as substantial dissolution of silica may occur in the water column as well as in the sediments, and the fraction of siliceous algae in the total productivity may also fluctuate [178]. Which species of diatoms produced the siliceous remnants is an important source of information. Diatoms differ in their preferred

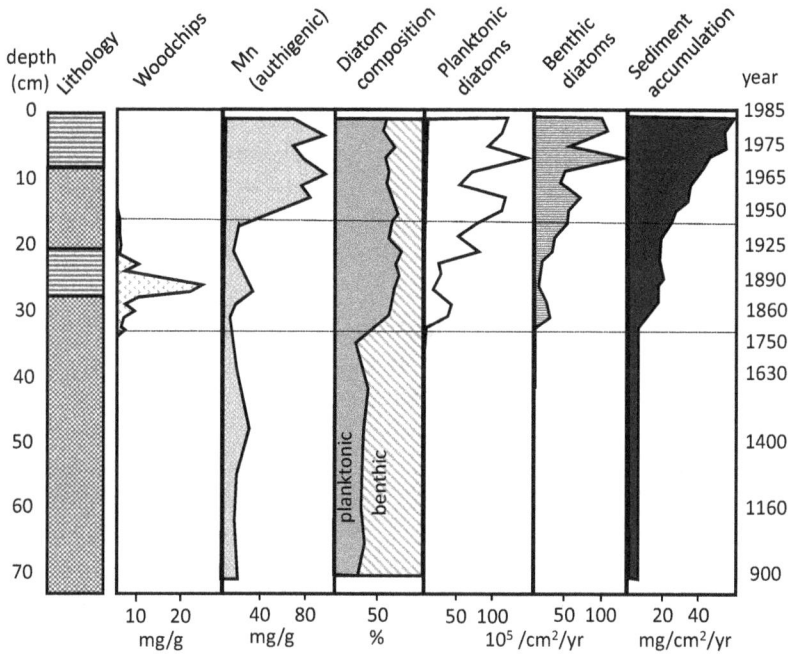

Fig. 10.4. The history of anthropogenic influences and environmental changes recorded by biomarkers and other sediment characteristics in a Vermont lake (Harvey's Lake; redrawn from [182]).

temperatures, salinities, and nutrient and light conditions, so changes in their relative abundances, recorded vertically within the sedimentary sequence, narrate the corresponding histories of changes in the environment.

10.2.3 Stable isotopes of carbon

Many biological and geochemical processes become recorded in the isotopic compositions of sediment substances. While different isotopes of the same chemical element have the same chemical properties, isotopes with smaller atomic masses tend to participate in reactions slightly more frequently than heavier isotopes: they diffuse faster when in the free state and also have higher vibrational frequencies when bound within molecules. This preferentially transfers lighter isotopes into the reaction's products, while leaving heavier isotopes behind in the unreacted fraction (see Appendix E).

Carbon has two stable isotopes: the "standard" ^{12}C and the less abundant ^{13}C. Different types of organic matter have different concentrations of ^{13}C isotopes relative to ^{12}C, which helps in identifying the source of the organic material. For example, the so-called C$_4$ land plants (notably, corn) are significantly enriched in the heavier carbon isotope relative to the more common C$_3$ plants. Organic materials from different sources also differ in their stoichiometric ratios of carbon to other major biological elements, such as nitrogen and phosphorus (Table D.1; Fig. 10.5).

Variations in the isotopic and elemental composition thus may record the changing sources of the organic material. These, in turn, are likely to reflect processes that took place in the water body or its watershed. They may also reflect changes in the rates of processes within the system. In particular, increases in lacustrine primary productivity may lead to increases in the δ^{13}C of organic matter. As during photosynthesis ^{12}C is incorporated from CO_2 into biomass preferentially, high levels of productivity deplete the CO_2 pool of lighter isotopes, causing a higher proportion of heavier ^{13}C to become incorporated during biosynthesis [186]. Conversely, the decomposition products of organic matter become isotopically light, i.e., depleted in the heavier ^{13}C isotope. In stratified water columns where isotopic compositions of organic material are affected

Table 10.1. Composition of sedimentary organic matter [3, 122, 124, 183–185].

Organic matter type	C/N	C/P	δ^{13}C
Vascular land plants	20–500	830 (300–1,300)	
C$_3$ land plants			−25 to −28‰
C$_4$ land plants			−8 to −18‰
Soil OM	8–15		−14 to −26‰
Marine phytoplankton	5–10	106–117	−17 to −22‰
Marine zooplankton	5–6	60–120	
Marine bacteria	3–5	7–80	
Marine seston		60–200	
Freshwater phytoplankton	6–20	115–775	−23 to −28‰
Freshwater zooplankton	5–6	80–230	
Freshwater bacteria	3–14	7–400	
Freshwater seston	7–20	20–800	

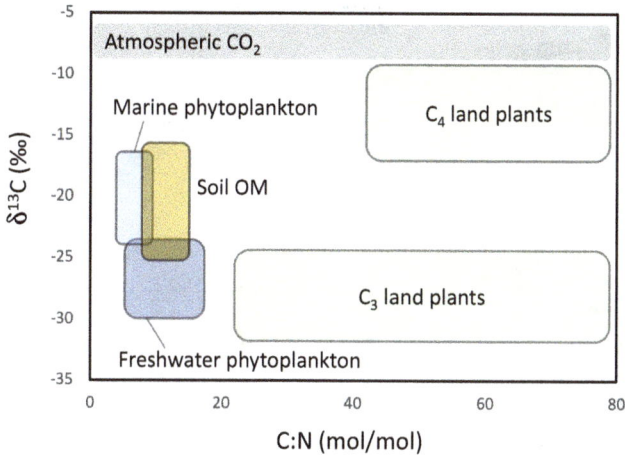

Fig. 10.5. Isotopic and elemental composition of common sources of organic carbon.

by multiple sources (both surface and deep) and their seasonal variations, the resultant effects can be quite complicated [187].

When used for reconstructing depositional histories, the profiles of ^{13}C need to be corrected for the *Suess effect*: a decrease in the atmospheric carbon pool $\delta^{13}C$ due to increased emissions of carbon from fossil fuels, which are ^{13}C-depleted [189] (Fig. 10.6). For plants and bacterial or algal phytoplankton that synthesize their biomasses from atmospheric CO_2, the isotopic composition of the produced fresh organic matter has thus varied over time, especially since the second half of the 20th century.

The stable isotopes of carbon are also frequently analyzed in carbonate ($CaCO_3$) phases, often together with oxygen isotopes ($^{18}O/^{16}O$) [190]. While carbon (as well as nitrogen) isotopes are used to infer changes in the carbon (and nutrient) cycling in the water body and its watershed, oxygen isotopes ($\delta^{18}O$) can help reconstruct past changes in temperature and the precipitation-evaporation balance [187]. Isotopic fractionations that become recorded in carbonates are generated by both equilibrium and kinetic effects (see Appendix E), which are imparted by different sets of processes. Equilibrium fractionations relate to the thermodynamics of mineral precipitation and can be used as "paleothermometers". For example, for carbonates precipitated under conditions near thermodynamic

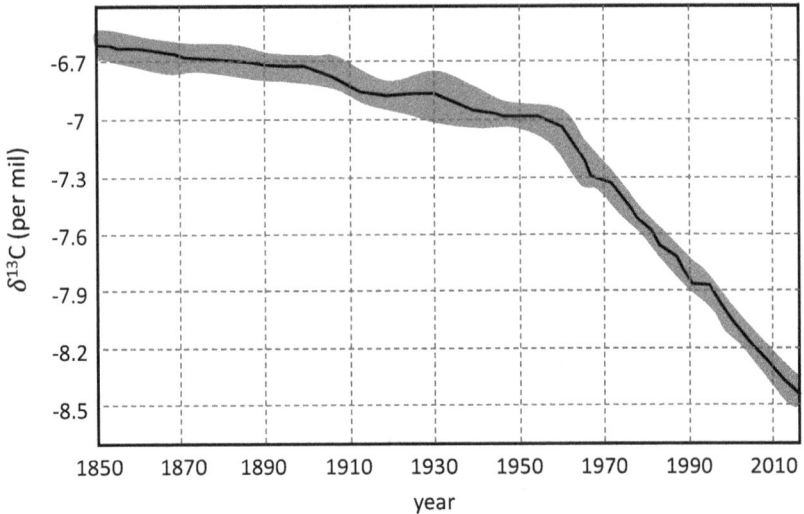

Fig. 10.6. The isotopic composition of atmospheric carbon dioxide (redrawn from [188]).

equilibrium, the $\delta^{18}O$ in the mineral decreases by about 0.24‰ for each 1°C increase in temperature [191]. In contrast, kinetic fractionations reflect conditions far from thermodynamic equilibrium and depend on the difference in rates at which the different isotopes are transferred during the reaction.

10.3 Diagenetic Alteration of Depositional Signals

Sediments alter depositional signals, passing them through a filter of focusing, bioturbation, organismal uptake, and biogeochemical transformations [178]. Stoichiometric ratios within organic matter can change, as certain elements are mineralized faster than others. Isotopic signals are modified by fractionations that occur during diagenesis. The elemental and isotopic compositions of substances within sediments thus reflect both the depositional signals and their overprint from diagenesis. So, in principle, the distributions of $\delta^{13}C$, C:N and C:P ratios, and other characteristics can provide clues to both the provenance and the diagenetic history of the material.

10.3.1 Carbon isotopes

As the diagenetic reactions that mineralize organic carbon are more likely to involve lighter isotopes, the resulting dissolved substances (DIC and DOC) become isotopically lighter than the source material. The isotopic composition of the produced dissolved pool thus reflects the extent of carbon mineralization in the sediment. The $\delta^{13}C$ of the DIC being exchanged with the overlying water column, for example, is often more negative in oxic sediments, where mineralization is more complete, than in anoxic sediments [187]. The solid-phase organic fraction remaining in the sediment correspondingly becomes progressively enriched in heavier isotopes during diagenesis, leading to higher $\delta^{13}C$ values in deeper sediment [192]. Seen from a historical perspective from older sediment to more recent, the $\delta^{13}C$ appears to become isotopically lighter with time. As this is the same direction as the Suess effect, separating the depositional and diagenetic isotopic signals requires quantitative comparisons [186]. The situation is further complicated by reservoir effects, such as when the organic material at the sediment surface is not fresh. Sediment transport and resuspension may cause the material near the sediment-water interface to be effectively older by several years, and sometimes by decades [193]. Accounting for this effect corresponds to a time correction for the Suess effect, often with non-trivial consequences, as illustrated in Fig. 10.7.

10.3.2 Vertical distributions of organic carbon

The organic carbon content of the sediment, to some degree, may be expected to track the history of its deposition, reflecting the ecosystem processes such as eutrophication. Historical records of carbon sedimentation, however, are overprinted by the diagenetic mineralization signal, which makes the concentrations of organic carbon decrease with depth into the sediment. The fraction of the deposited organic carbon that becomes mineralized depends on the conditions within the sediment and can be affected by factors such as the degree of sediment oxygenation, the intensity of sediment reworking by currents and biota, and the nature of the organic material itself (e.g., terrestrial vs. aquatic). For this reason, paleoceanographic and

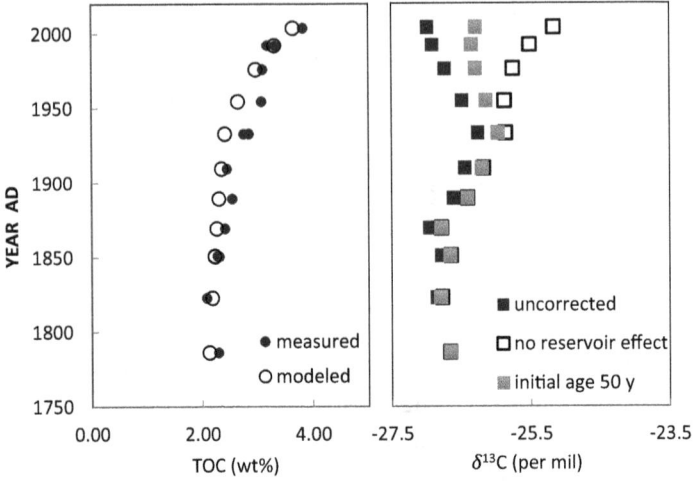

Fig. 10.7. Illustrated effect of applying the ^{13}C Suess correction with and without the reservoir effect (Lake Superior sediment, core BH09-3 of [186]).

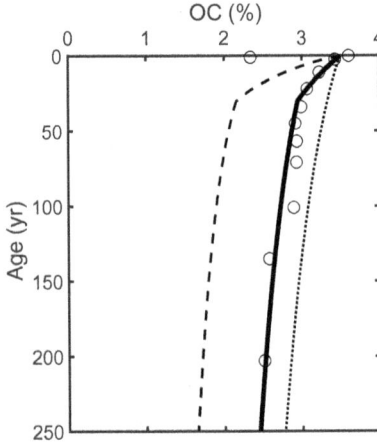

Fig. 10.8. Calculated profiles of organic carbon, illustrating the effects of oxygen exposure and the freshness of the deposited organic matter. The profiles were calculated from Eqs. 11.27–11.29. Solid line: the oxygen exposure time is 30 years, the initial carbon age is 80 years. Dashed line: the initial carbon age is 20 years. Dotted line: no oxygen exposure in sediment. Symbols are data from a deeply (3.5 cm) oxygenated sediment in Lake Superior [72].

paleolimnological reconstructions need to rely on additional information, such as from stable carbon isotopes and biomarkers. Nevertheless, knowing the approximate rate law for carbon mineralization in sediments makes it possible to derive at least an approximate shape of the organic carbon distribution that would result from a steady deposition and mineralization under hypothesized conditions (Fig. 10.8; see Chapter 11 for the derivation of such steady state solutions). Significant deviations from such profiles could reflect the historical trends in deposition.

10.4   Exercises

1. If the unsupported activity of ^{210}Pb decreases by a factor of two every 3 cm inside the sediment, what is the approximate burial velocity? (Assume constant porosity.)
2. If, in the previous exercise, the porosity is not constant but changes from 0.98 to 0.96 over the depth interval where the decay activity halves, what is the burial velocity for that interval?
3. What is the maximum possible magnitude of the Suess effect? In what ways could the data be misinterpreted if the Suess effect is not taken into consideration?
4. In sediment A, the concentration of particulate organic carbon (as wt%) is found to *increase* from the sediment-water interface downward. What can be safely concluded about this sediment and the history of its environment? In sediment B, the POC concentration is found to *decrease* from the sediment-water interface downward. What can be safely concluded about this sediment and the history of its environment? If different interpretations are possible in either case, what types of analyses could be performed to resolve the ambiguity?

Chapter 11

Mathematical Descriptions of Diagenesis

Sediment geochemistry is viewed by many as an empirical science that is carried out almost exclusively by measurements and experiments. This view misses an immense arsenal of tools that have been responsible for many significant advances in the discipline. To limit a textbook on sediment geochemistry solely to its phenomenological aspects would be like trying to describe planetary motion by relying only on Kepler while not mentioning Newton. Summarized first by Robert Berner [1], the theoretical approach to early diagenesis enables rigorous, quantitative analyses of processes that would otherwise be evaluated only qualitatively. It helps make sense of datasets, deduce effects that are intractable by experiment alone, and extrapolate beyond the observational data. Much of the theory presented here was used with good success in constructing numerical diagenetic models. The goal of this chapter extends well beyond numerical applications, however. The mathematical fundamentals presented here may be used on their own, without numerical modeling, in many investigations of sedimentary processes. The obtained theoretical solutions provide a formalism that lends a universal context to many site-specific data.

11.1   The Diagenetic Equation

As illustrated throughout this book, diagenesis is shaped by tightly coupled interactions among a plethora of physical, chemical, and biological processes. The complexity of these interactions, however, can be neatly captured by a single equation, which can be solved for the chemical species of interest. It is commonly referred to simply as *the diagenetic equation*.

11.1.1   Derivation

Consider a volume of sediment of thickness dz with a cross-sectional area A (Fig. 11.1). Substances can be transported vertically through the top and bottom boundaries of this volume by diffusive fluxes F. The concentration of a chemical species $C(z, t)$ within that volume may change in time for one or more of the following reasons:

(a) If the diffusive fluxes of the substance through the top and bottom boundaries are different, resulting in the accumulation (or consumption) of the substance within the volume. The difference between the fluxes $(\mathrm{mol\,cm^{-2}s^{-1}})$ in and out of the volume gives the amount (Q, mol) of substance that accumulates within the volume per unit time:

$$\frac{dQ}{dt} = [F(z) - F(z + dz)]A \qquad (11.1)$$

As this amount is distributed over the volume $dV = A \cdot dz$, the rate $(\mathrm{mol\,cm^{-3}\,s^{-1}})$ at which the concentration would increase

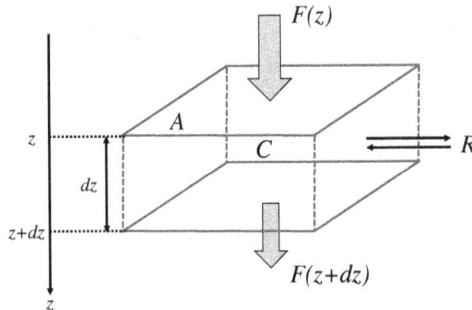

Fig. 11.1.   Conceptual illustration for the derivation of the diagenetic equation. The vertical fluxes F in and out of a horizontal layer of sediment, and the rates R within the layer, affect the concentrations C within it.

(or decrease) is

$$\left(\frac{\partial C}{\partial t}\right)_{\text{diff}} = \frac{F(z) - F(z+dz)}{dz} = -\frac{\partial F}{\partial z} \qquad (11.2)$$

The notation ∂, for partial derivative, is used here because the concentration C and the flux F may be functions of both time t and space z. The partial derivative is taken with respect to only one of those variables. Using the Fick's law of diffusion (Eq. (2.10)), the last expression can be written using the diffusion coefficient D:

$$\left(\frac{\partial C}{\partial t}\right)_{\text{diff}} = \frac{\partial}{\partial z}\left(D\frac{\partial C}{\partial z}\right) \qquad (11.3)$$

(b) If the advective fluxes through the top and bottom boundaries are different. Similarly to the treatment above, the effected increase in concentration can be calculated from the difference in the advective fluxes:

$$\left(\frac{\partial C}{\partial t}\right)_{\text{adv}} = \frac{vC(z) - vC(z+dz)}{dz} = -\frac{\partial(vC)}{\partial z} \qquad (11.4)$$

(c) If the substance is produced or consumed within the volume by a reaction. The concentration C in this case changes in time at the net (combined) rate R ($\text{mol cm}^{-3}\text{s}^{-1}$) of all the reactions that consume or produce the substance.

$$\left(\frac{\partial C}{\partial t}\right)_{\text{reac}} = R_{\text{net}} \qquad (11.5)$$

(d) If the substance is brought into (or removed from) the volume by some other (non-local) process that bypasses the cross-sectional boundaries, such as when it is injected through a penetrating animal burrow. The rate of accumulation in this case depends on the rate of injection (see below).

$$\left(\frac{\partial C}{\partial t}\right)_{\text{nonloc}} - R_{\text{nonloc}} \qquad (11.6)$$

As sediment reactions occur either in the porewater or the solid matrix, the corresponding concentrations need to be adjusted for porosity. When the concentrations of dissolved species are expressed per volume of porewater $(\mathrm{mol/cm^3_{pw}})$ rather than per total sediment volume $(\mathrm{mol/cm^3_{sed}})$, they need to be multiplied by the porosity ϕ $(\mathrm{cm_{pw}/cm_{sed}})$. Concentrations of solid-phase species, accordingly, need to be multiplied by the corresponding volume fraction for the solid phase $(\mathrm{cm_{sol}/cm_{sed}})$: $\phi_s = 1 - \phi$.

11.1.2   The full equation

Combining the terms for the individual processes (Eqs. (11.3)–(11.6)), we obtain the *diagenetic equation* that governs the evolution of chemical concentrations $C(z,t)$ in space and time. For dissolved species, the equation is

$$\frac{\partial(\phi C)}{\partial t} = \frac{\partial}{\partial z}\left(\phi D_m \frac{\partial C}{\partial z}\right) - \frac{\partial}{\partial z}(\phi v C) + \phi R_{irr}(z) + \sum_i \phi \nu_i R_i$$

(11.7)

The diffusion coefficient D_m for solutes is the molecular diffusion coefficient (Table 2.1), with a correction for sediment tortuosity (Eq. (2.11)). ν_i are the stoichiometric coefficients for each reaction R_i in which the species of interest participates. The term $R_{irr}(z)$ describes the intensity of non-local processes such as bioirrigation (see Section 11.3.7).

For solid-phase species, the corresponding equation is

$$\frac{\partial(\phi_s C)}{\partial t} = \frac{\partial}{\partial z}\left(\phi_s D_b \frac{\partial C}{\partial z}\right) - \frac{\partial}{\partial z}(\phi_s v C) + \sum_i \phi_s \nu_i R_i \qquad (11.8)$$

The diffusion coefficient for solid phases D_b is the bioturbation coefficient. The solid volume fraction ϕ_s is the equivalent of porosity for solid species: $\phi_s = 1 - \phi$, i.e., the fraction of the overall sediment volume occupied by the solid fraction $(\mathrm{cm^3_{solid}/cm^3_{sed}})$.

11.1.3   Boundary and initial conditions

As for any differential equation where variables can evolve in space and time, specific solutions of the diagenetic equation would depend on the choice of the *initial* and *boundary* conditions. To solve for the concentrations $C(z,t)$, one needs to specify the vertical distributions $C(z,t_0)$ at some initial moment of time t_0. One also needs to specify how the profiles $C(z,t)$ are constrained at the upper and lower boundaries of the domain: e.g., at the sediment surface $z = 0$ and at the lower boundary of the diagenetically active layer $z = L$. For the second-order differential equation, these latter boundary conditions can be of two types. They can specify at each boundary the concentrations themselves: $C(0,t) = C_0(t)$ (the *Dirichlet*-type boundary condition); or, they can specify the steepness of the spatial gradients of the concentrations: $\left(\frac{\partial C(z,t)}{\partial z}\right)_{z=0} = \left(\frac{\partial C}{\partial z}\right)_0 (t)$ (the *Neumann*-type boundary condition). The upper and lower boundaries for the same concentration profile may have different types of boundary conditions.

For instance, if the water overlying the sediment remains well-oxygenated at all times, the boundary condition at the sediment-water interface for the oxygen concentration profile may be chosen to be of the fixed-concentration type: $C_{O2}(0,t) = C_{O2}^0$. The gradient of oxygen across the interface (and hence the diffusive flux and the sediment's oxygen uptake) is then not explicitly imposed and can be obtained as the diagenetic equation's solution that reflects the conditions inside the sediment. In the deep sediment, however, the reactivity of organic matter may be considered sufficiently low for the oxygen consumption rate to be negligible, so that the underlying sediment layers exert no oxygen demand and the diffusion of oxygen would eliminate any concentration gradients. The boundary condition at the lower boundary of the domain, in that case, could be chosen as a no-flux condition: $\left(\frac{\partial C_{O2}}{\partial z}\right)_{z=L} = 0$. This condition of no gradient (corresponding to a flat oxygen profile at its bottom end) would hold regardless of whether oxygen is actually present in the deep sediment, i.e., the oxygen concentration is not explicitly constrained and can be obtained as the equation's solution. In practice, the initial and boundary conditions are typically chosen depending on the type of the sedimentary environment being described and on the type of the problem at hand.

11.2   Approximations and Simple Solutions

Though the diagenetic Eqs. (11.7) and (11.8) may look complicated, in many instances they can be substantially simplified by omitting terms whose magnitudes are negligible in the environment of interest. Evaluating the relative magnitudes of the terms is often possible by quick order-of-magnitude estimates (see Section 2.5). For example, in non-permeable (muddy) sediments, the advection term reflects only the downward burial motion and in the upper sediment layer may be neglected for solutes, as it is small in comparison to the molecular diffusion term. Porewaters are being transported downward by burial by a few millimeters per year, whereas, on the scale of centimeters, molecular diffusion covers the same distance in several hours (Table 2.2). Likewise, the diffusive term can be neglected for solid species below the zone of bioturbation. A crucial simplification can be achieved, however, if one can neglect the time derivative term, in a so-called steady-state approximation.

11.2.1   The steady state approximation

When environmental conditions remain approximately the same over a long time, the concentration profiles of chemical species no longer vary in time. In that case, the derivative with respect to time at the left-hand side of the diagenetic equation can be set to zero, resulting in a highly useful *steady state approximation*. As the concentrations $C(z,t)$ are now only the functions of depth, $C(z)$, the diagenetic equation can be written in ordinary, rather than partial, derivatives. For solutes (neglecting advection relative to molecular diffusion and omitting bioirrigation for simplicity), this gives

$$0 = \frac{d}{dz}\left(\phi D_m \frac{dC}{dz}\right) + \sum_i \phi \nu_i R_i(z) \tag{11.9}$$

For solids

$$0 = \frac{d}{dz}\left(\phi_s D_b \frac{dC}{dz}\right) - \frac{d}{dz}(\phi_s v C) + \sum_i \phi_s \nu_i R_i(z) \tag{11.10}$$

An important consequence of steady state is that conditions at different depths within the sediment become linked by the conservation of mass. As no substance can accumulate or disappear within the sediment (otherwise its concentration would have to increase or decrease with time), vertical fluxes at any two depths must be coupled: Whatever substance moves vertically at depth z_1 must either move at the same flux at any other depth z_2, or the difference needs to be accounted for by the rates of chemical reactions in the depth interval between z_1 and z_2. The same constraint links the top and bottom of the diagenetically active zone. Organic carbon deposited at the sediment surface must be either buried into the deep sediment (e.g. as organic matter or carbonate minerals) or be mineralized and removed back to the water column as CO_2, methane, or DOM. Phosphorus deposited to the sediment with particulate organic matter must be either buried into the deep sediment (as organic or mineral P) or recycled back to the water column as dissolved phosphate – irrespective of the processes within the sediment, however complex they may be. And so on. This gives a powerful way of characterizing some of the most important diagenetic fluxes and rates without needing to know the intricacies of biogeochemical cycling.

Test your understanding: Phosphorus is deposited to the sediment surface with particulate organic matter (POM); the steady state sedimentation flux is 0.1 mmolP m^{-2} y^{-1}. The burial efficiency of organic matter into the deep sediment is 10%. No P-bearing minerals are detected in the sediment below the diagenetically active zone. Which biogeochemical processes within the sediment need to be characterized to predict the phosphate flux from sediment into the overlying water column? (The answer is on page 194.)

Estimating fluxes and rates

The steady state approximation provides a convenient relationship between the vertical fluxes F of chemical species and their respective volume-specific rates R [11, 31]. Integrating Eq. (11.9) over depth from some depth z_0 to z gives

$$F_z = F_{z_0} + \int_{z_0}^{z} \sum_i \phi \nu_i R_i \, dz = F_{z_0} + \int_{z_0}^{z} \phi R_{net} \, dz \qquad (11.11)$$

where $F = -D_m \frac{dC}{dz}$ is the (downward) molecular diffusion flux of the solute. The change in the flux of the substance in the depth interval from z_0 to z is thus equivalent to the depth-integrated net rate (R_{net})

of all the chemical reactions in which that substance participates. For example, the downward flux of oxygen at the sediment surface is equivalent to the depth-integrated rate of all the reactions that consume oxygen in the sediment. As discussed in Chapter 5, the latter is a good approximation for the overall rate of carbon mineralization.

Likewise, for solid substances, the downward flux due to bioturbation and burial is

$$F = -\phi_s D_b \frac{dC}{dz} + \phi_s v C$$

Integrating Eq. (11.10) over depth, one obtains

$$F_z = F_{z_0} + \int_{z_0}^{z} \phi_s R_{net}\, dz \qquad (11.12)$$

Similarly to the Eq. (11.11) for solutes, fluxes change from one sediment depth to the next by the amount that is equivalent to the depth-integrated net rate. For example, for particulate organic carbon (POC), if z is taken at the bottom of the diagenetically active zone, the difference between the fluxes of organic carbon at the sediment-water interface (F_{z_0}) and the deep sediment (F_z) gives the extent of the overall sediment respiration.

Example: Oxygen consumption rates.   Oxygen diffusing from overlying water is consumed in the upper centimeters of sediments. Choosing z_0 in Eq. (11.11) to be below the depth of oxygen penetration so that $F_{O_2}(z_0) = 0$, Eq. (11.11) becomes

$$F_{O_2}(z) = \int_{\infty}^{z} \phi(z') R_{net}(z')\, dz' \qquad (11.13)$$

That is, at steady state, the downward fluxes of oxygen at depth z equal the depth-integrated rates of oxygen consumption below that depth. By differentiating this equation with respect to z, the volume-specific rates of oxygen consumption R_{net} ($mol\, cm^{-3}\, d^{-1}$) equal the derivative of the oxygen flux with respect to depth:

$$\phi R_{O_2} = -\frac{dF_{O_2}}{dz} \approx \frac{d}{dz}\left(\phi D_{O_2} \frac{d[O_2]}{dz}\right) \qquad (11.14)$$

(The same result for diffusive fluxes can be obtained directly from Eq. (11.9).) The meaning of this relationship is straightforward:

when the downward oxygen flux changes from one sediment layer to another, it is because oxygen is consumed within the layer. As discussed earlier, except in a narrow depth interval around the OPD where oxygen is consumed by the reduced products of anaerobic metabolisms diffusing from deeper sediment, the rate R_{O_2} is approximately equal to the rate of carbon mineralization (R_G) at the same depth. Using this equation, the rate of organic matter mineralization then can be calculated from high-resolution oxygen data, such as from experimentally obtained profiles of oxygen in the sediment porewater.

Test your understanding: The concentration of methane in a marine sediment at steady state increases with depth below the sulfate-methane transition zone (SMTZ). The concentration rises approximately linearly with depth below the SMTZ, from practically zero at the SMTZ to 0.5 mM at the depth of 50 cm below it. In deeper sediment, the concentration continues to increase, but at a progressively slower pace, and the profile becomes essentially flat 3 m below the SMTZ. Is it possible to estimate the depth-integrated rate of methanogenesis $(\text{mmol}\,\text{m}^{-2}\,\text{y}^{-1})$ in this sediment? (The answer is on page 195.)

11.2.2 Simple illustration of sources and sinks

Let us illustrate how sources and sinks shape the vertical profiles of chemical species in sediments. Consider a steady state situation, for simplicity. The shape of the vertical profile of some dissolved species $C(z)$ in this case is described by Eq. (11.9) and its associated boundary conditions. For clarity, we will write it here for a case of constant porosity ϕ and with a single reaction term R:

$$D_m \frac{d^2 C}{dz^2} = -R(z) \tag{11.15}$$

Just as before, $R(z)$ here is positive when the species C is being produced (a source). The vertical profile $C(z)$ then can take the following shapes.

Case 1 No reaction. In the absence of either sinks or sources $(R(z) = 0)$, the second derivative of $C(z)$ is zero, which means that the first derivative is constant and the profile has to be a straight line. (The general solution of Eq. (11.15) is $C(z) = const_1 + const_2 \cdot z = a + bz$, which is an equation of a straight line with slope b and intercept a.) The slope of the line (the first derivative) is the same throughout the domain, which means that the diffusive fluxes are the same. This is a natural consequence of the absence of sources or sinks: what enters the domain at the top must exit at the bottom. Accordingly, straight vertical profiles $C(z)$ indicate the absence of reactions. A non-zero slope of the line (i.e., the concentration at the top of the domain is different than at the bottom) signifies a non-zero flux. As at steady state the substance cannot accumulate within the domain, a (say) downward flux of the substance should mean that the substance is supplied at the top and removed at the bottom (either by physical transport or by a reaction beyond the boundary of the domain).

Case 2 Constant sink or source. A sink or a source of a substance is going to change the slope of its steady state profile, modifying the diffusive flux $F(z) = -D_m \frac{dC}{dz}$. Integration of Eq. (11.15) for a constant rate $(R(z) = const)$ leads to

$$F(z) = F_{z=0} + Rz \tag{11.16}$$

$$C(z) = C_{z=0} + \left. \frac{dC}{dz} \right|_{z=0} z - \frac{R}{D_m} z^2 \tag{11.17}$$

For example, if the substance is being consumed $(R < 0)$ at a constant rate throughout the domain, its downward flux $F(z)$ decreases with depth. The concentration profile is a concave parabola (negative second derivative). If the substance is being produced $(R > 0)$, the concentration profile is a convex parabola (positive second derivative). Note that Eq. (11.16) allows the flux $F(z)$ to switch sign within the domain: it may start negative (i.e., upwards) at the top of the domain, then become zero within the domain, then become positive (downward) in the deeper sediment. The slope of the profile thus would also switch sign. The curvature of the profile (concave or convex), however, would remain the same. The variety of possibilities afforded by Eqs. (11.16)–(11.17) are illustrated in Fig. 11.2.

While covering only a simple case of constant rates, Fig. 11.2 serves as a useful guide to interpreting the vertical profiles of chemical

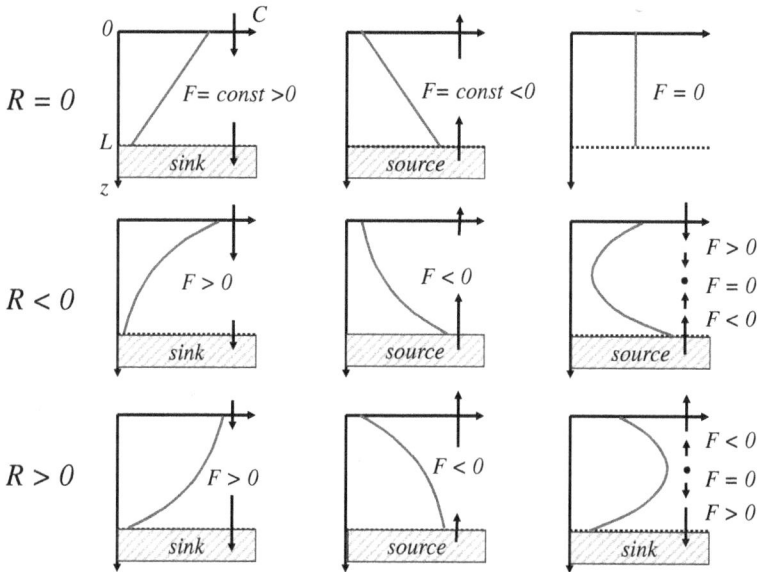

Fig. 11.2. Profile shapes generated by positive or negative reaction rates.

species: A peak in a steady-state profile indicates the substance is being produced (a positive net rate). A dip in the profile indicates the substance is being consumed (a negative net rate). A profile where the negative slope becomes flatter, or the positive slope becomes steeper, indicates consumption. A profile where the positive slope becomes flatter, or the negative slope becomes steeper, indicates production. A straight-line profile indicates the absence of sources or sinks, except perhaps at the either end of the line.

Furthermore, a sustained peak in a steady-state concentration profile means not only that the substance is being produced near the peak, but also that it is being removed (either by transport or reaction) both above and below the peak. Otherwise, the diffusion would eventually flatten any concentration gradient. The same applies, of course, to a sustained dip in a steady state profile.

It follows from the above discussion that a switch in the convexity of the profile should indicate a transition between the consumption and production zones within the sediment, such as those that exist for many redox-sensitive species. For example, ammonium is net-produced in the anoxic sediment through the decomposition of

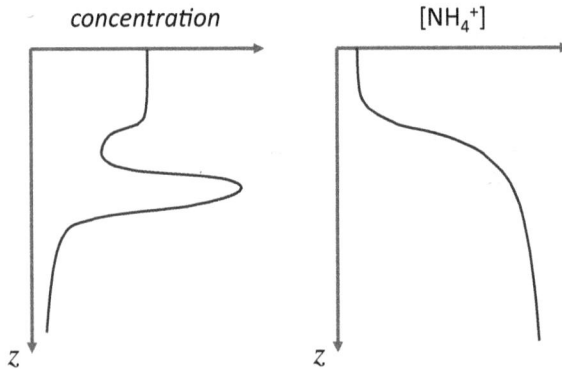

Fig. 11.3. Test your understanding: where is each substance being produced or consumed?

organic matter, and net-consumed in the oxic zone via oxic nitrification. The inflection point in the profile then should indicate the boundary between the net-production and net-consumption.

Test your understanding: Figure 11.3 shows some steady state concentration profiles. Indicate the sediment regions where the substance is being produced, consumed, or where its net reaction rate is zero.

11.2.3 Non-steady state diagenesis

In non-steady state, one useful solution of the diagenetic equation is the one that describes diffusion of a non-reactive tracer. For instance, in incubation experiments, a tracer such as Br^- is introduced into the water overlying the sediment and is allowed to diffuse into the sediment where its concentration is initially zero. When the reaction term in Eq. (11.7) is zero (and porosity is assumed constant), such a diffusive problem is described by the differential equation:

$$\frac{\partial C}{\partial t} = \frac{\partial}{\partial z}\left(D_m \frac{\partial C}{\partial z}\right) \qquad (11.18)$$

When the boundary condition is chosen so that the concentration at the sediment-water interface is maintained constant at C_0, the equation effectively describes diffusion of a tracer from a large inexhaustible reservoir. The solution to this equation is a so-called complementary error function *erfc* [194]:

$$C(z,t) = C_0 \operatorname{erfc}\left(\frac{z}{2\sqrt{D_m t}}\right) \qquad (11.19)$$

The *erfc* function is included in most software packages. The solutions are illustrated in Fig. 11.4. Note that the characteristic length scale for the penetration of the tracer increases with time nonlinearly, as $\sqrt{D_m t}$, i.e., the downward propagation of the tracer slows down with time. The relationship given by the above equation allows one to estimate the typical time scales for the experiment, as well as compare the observed profiles to the theoretical ones. The profiles that are due to purely molecular diffusion, as shown in Fig. 11.4, indicate the minimum expected penetration of the tracer into the sediment. Enhanced downward transport of the tracer can be used to characterize the intensity of other transport processes, such as bioirrigation [195].

The diffusive boundary layer, which may extend several millimeters above the sediment-water interface, may create a gradient in the tracer concentration, resulting in the concentration at the interface being different from the concentration in the bulk reservoir.

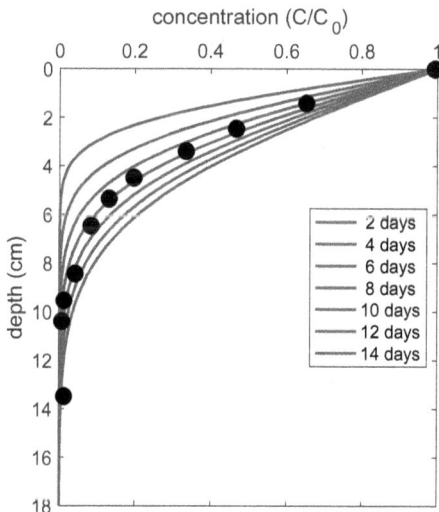

Fig. 11.4. Theoretical and observed solutions for bromide diffusion in a tracer addition experiment. Curves are from Eq. (11.19). Data points are from a bromide tracer incubation experiment of [195], for a marine site with minimal bioirrigation.

To avoid this, the overlying water is typically being stirred during incubations.

11.3 Rate Formulations

11.3.1 Organic carbon mineralization

Decay of natural organic matter in sediments is a complex biogeochemical process, but a number of good approximations have been suggested to describe its rates in different environments. Probably the simplest way to represent this process is to imagine that organic matter possesses some fixed "reactivity" k. A fixed reactivity implies that in each time interval some fixed proportion of the organic molecules decay and become mineralized. This representation, like for a radioactive decay, leads to a first-order kinetic law: the rate of decay (the number of organic carbon atoms mineralized per second) is proportional to the amount (or concentration) of the organic material:

$$R_G = \frac{dG}{dt} = -kG \qquad (11.20)$$

Here, we use the commonly used notation where the concentration of organic matter is denoted with letter G. The minus sign corresponds to the fact that the concentration G decreases with time, i.e., the rate R_G is negative (meaning consumption).

   This formulation is known as the $1G$ *model*. It assumes that all of the sedimentary organic matter can be represented by a single pool G, characterized by a fixed reactivity k. Anyone familiar with ordinary differential equations probably recognizes that any equation like Eq. (11.20) where the rate of change is proportional to the quantity itself is a recipe for an exponential solution. Integrating both sides of the equation with respect to time gives that the amount of organic material decreases with time exponentially:

$$G(t) = G_0 e^{-kt} \qquad (11.21)$$

where G_0 is the initial amount of organic matter. The reactivity k defines the speed of this exponential decay and has the units of inverse time (e.g., y^{-1}). The inverse quantity, $\tau = 1/k$, defines the time scale over which the amount of organic matter decreases by a factor of $e = 2.71828....$ A more easily imagined quantity, the time over which the amount of organic matter decreases by a factor of 2 (the "half-life" in the parlance of radioactivity), can be obtained as

$$\tau_{1/2} = \frac{1}{k} \ln 2 \approx \frac{0.69}{k} \qquad (11.22)$$

For example, reactivity of $k = 2$ y^{-1} means that the exponential time scale for the decay is 0.5 years, and that only half of the organic material will remain after $0.69/2 = 0.35$ years. Figure 11.5 illustrates the single-exponential decay solutions for different values of k.

It is fairly obvious that representing the entire pool of organic matter as a single homogeneous mass that decays at the same constant rate is rather unrealistic. Real sediments retain substantial quantities of organic material even after prolonged diagenesis, oil and coal being good testaments to this geological fact. It is obvious that some fractions of the organic pool decay faster than others.

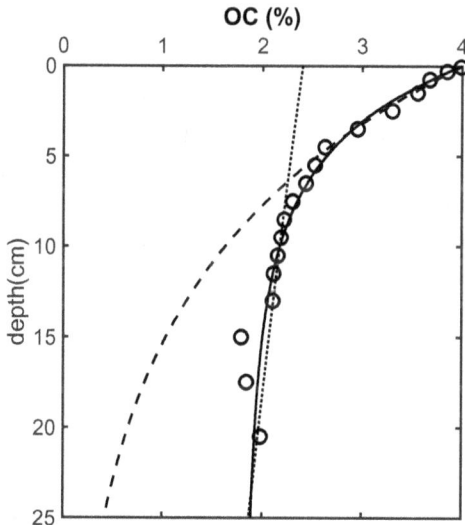

Fig. 11.5. Fitting of a vertical profile of organic carbon by a 1G model (for high (dashed) and low (dotted) reactivities k) and a 2G model (solid line). For simplicity, a linear relationship is assumed between depth z and time t, i.e., a constant burial velocity. The data points are from [11] for an offshore sediment in Lake Superior.

For example, sugars are easily used by microbes, whereas cellulose and lignins are notably recalcitrant.

A *2G model* provides a step-up in realism by splitting the organic matter into two pools: "reactive" and "refractory". Each of the pools are then assumed to obey the same first-order kinetics as in the 1G model, but with different reactivities:

$$R_G = \frac{d(G_1 + G_2)}{dt} = \frac{dG_1}{dt} + \frac{dG_2}{dt} = -k_1 G_1 - k_2 G_2 \qquad (11.23)$$

The double-exponential dynamics then becomes

$$G(t) = G_1^0 e^{-k_1 t} + G_2^0 e^{-k_2 t} \qquad (11.24)$$

It provides a noticeable improvement over the 1G model (Figure 11.5). The values of the two reactivity constants, k_1 and k_2, are typically on the order of $1 \ \mathrm{y}^{-1}$ for the labile fraction and $0.01 \ \mathrm{y}^{-1}$ for the refractory one (compare to the corresponding reactivities k in Fig. 5.6). While still simplistic, Eq. (11.23) can be used to capture the profiles of organic carbon in sediments with some realism. Converting time t into depth z requires knowledge of the sedimentation history, and gets complicated if bioturbation is present. In simple cases, however, even a straightforward substitution of $t = z/v$, where v is a constant burial velocity, can yield satisfactory results (Figure 11.5). Alternatively, depth profiles $G(z)$ can be calculated from time dependences $G(t)$ when burial velocities (or sedimentation rates) are known. The depth within the sediment that the organic particles reach in time t (neglecting bioturbation) is

$$z(t) = \int_0^t v(t) dt \qquad (11.25)$$

Or, the time corresponding to a particular depth z within the sediment is

$$t(z) = \int_0^z \frac{dz}{v(z)} \qquad (11.26)$$

where the burial velocity $v(z)$ can be calculated from Eq. (2.4).

Further refinements along the same path result in a family of "multi-G models", with greater numbers of organic pools. The limiting case, with a continuous spectrum of the infinite number of reactive types, is known as the *reactive continuum* approximation [196].

An alternative (but ultimately equivalent [70]) approach is to consider all of the organic matter as a single pool whose overall bulk reactivity k decreases with time, as it progressively loses its most reactive fractions.

The phenomenological (with some theoretical backing [197]) power law of organic carbon reactivity given by Eq. (5.4) predicts a slower-than-exponential decay for organic carbon. The shape of the resulting vertical profile of organic carbon can be obtained by integrating the power law over time [72]. For either oxic or anoxic mineralization, the organic carbon concentration decreases with time as

$$C(t) = C(t_0)\exp\left[-\frac{b(t^{1-a} - t_0^{1-a})}{1-a}\right] \quad (11.27)$$

Here, t_0 is the age of the organic material when it reaches the sediment surface (typically days to months, but may be much longer if sediment resuspension and lateral transport along the bottom are substantial).

When differences between the mineralization rates in oxic vs. anoxic environments are taken into account, the vertical distribution of organic carbon within the oxic zone may be calculated (again, ignoring bioturbation!) from the same Eq. (11.27) using the corresponding values of the parameters a and b for oxic mineralization. The concentration at the bottom of the oxic zone, C_{OPD}, can be calculated from Eq. (11.27) as

$$C_{\text{OPD}} = C_0 \exp\left[-\frac{b_{\text{oxic}}(t_{\text{OPD}}^{1-a_{\text{oxic}}} - t_0^{1-a_{\text{oxic}}})}{1-a_{\text{oxic}}}\right] \quad (11.28)$$

The time t_{OPD} here includes the time that the organic material is exposed to oxygen in the sediment, as well as the time t_0 it took to arrive to the sediment surface by settling through the water column. The initial concentration C_0 is the concentration of organic carbon at the sediment-water interface.

The concentration of organic carbon below the depth of oxygen penetration (OPD) then can be calculated [72] as

$$C = C_{\text{OPD}} \exp\left[-\frac{b_{\text{anox}}}{1-a_{\text{anox}}}(t^{1-a_{\text{anox}}} - t_{\text{OPD}}^{1-a_{\text{anox}}})\right] \quad (11.29)$$

where a_{anox} and b_{anox} are the corresponding parameters for anaerobic mineralization in the reactivity law (Eq. (5.4)). Examples of

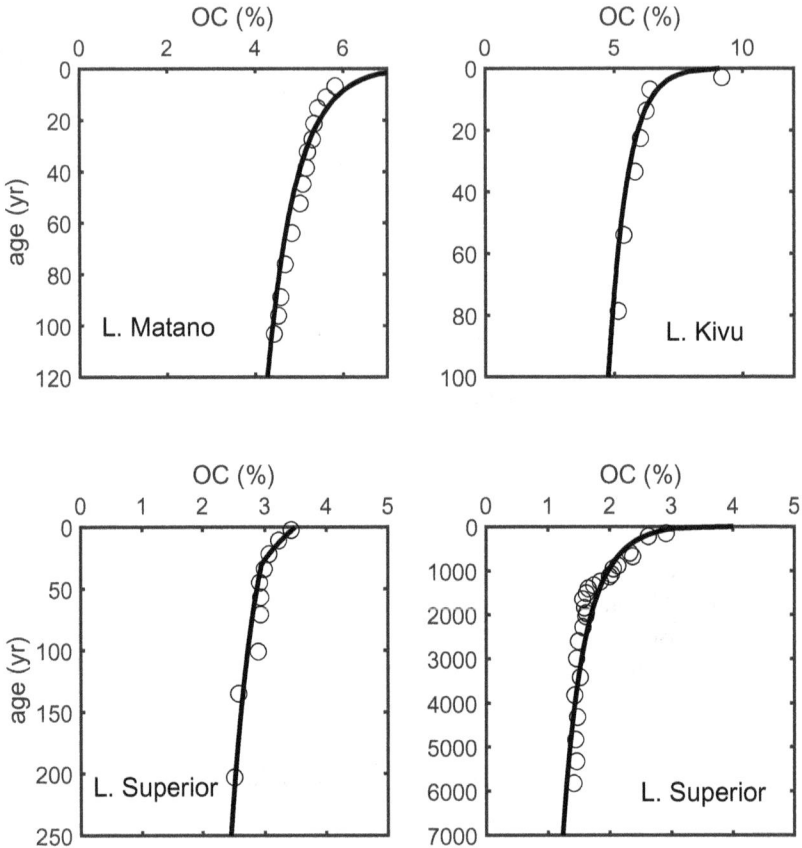

Fig. 11.6. Fitting of the vertical profiles of organic carbon (OC) by a power-law model with separate expressions for oxic and anoxic mineralization. The sampling sites span a range of bottom water oxygen concentrations: low oxygen in Lake Matano, anoxic (euxinic) in Lake Kivu, and fully oxygenated in Lake Superior. The two profiles from Lake Superior correspond to samples taken with a short (multicorer) core and a long (piston) core, hence the difference in time scales. (Adapted from [72].)

calculations using these equations are given in Figs. 10.8 and 11.6, and [72, 186].

11.3.2 Microbially-catalyzed reactions

Microbially-mediated reactions, such as those in the redox sequence of the organic matter mineralization, are commonly described using the so-called Monod kinetics:

$$R = V_{\max} \frac{C}{C + K_m} \tag{11.30}$$

Here, R is the reaction rate (e.g., $\mathrm{mol}\,L^{-1}h^{-1}$), C is the substrate concentration, $V_{\max}$ ($\mathrm{mol}\,L^{-1}h^{-1}$) is the reaction rate constant, and K_m is the "half-saturation" constant, which describes the microbe's affinity for the substrate. The kinetic parameter $V_{\max}$ corresponds to the maximum rate of the reaction when the substrate is abundant ($C \gg K_m$). For example, for the primary redox reactions in the organic carbon mineralization sequence, it corresponds to the maximum rate of organic carbon mineralization: $R_G = kG$ (Eq. (11.20)). The affinity constant K_m corresponds to the concentration of the substrate at which the rate is half of the maximum rate.

The Monod equation (Eq. (11.30)) is phenomenological, chosen to have a functional form that mimics the relationships that are often observed in experiments. It is, however, closely connected to the Michaelis–Menten equation, which is used to describe catalytic chemical reactions. This similarity is rooted in the mechanisms by which microbes catalyze geochemical reactions with their enzymes. Enzymes are proteins that catalyze reactions. The enzymes (E) bind with reactive substrates (S), forming intermediate complexes from which the substrates can be transformed into the reaction products (P):

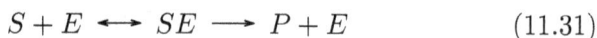

$$S + E \longleftrightarrow SE \longrightarrow P + E \tag{11.31}$$

The intermediate complexes lower the activation barrier, which would otherwise prevent the reaction from happening. The enzyme itself is not consumed during the reaction, acting only as a catalyst. The Michaelis–Menten equation describes the forward reaction rate for an enzymatic reaction (Eq. (11.31)), and its functional form coincides with Eq. (11.30) [5]. The mechanism of enzymatic catalysis suggests that the rate constant $V_{\max}$ should depend on the total concentration of the enzyme $[E]$, which is approximately proportional to the total number of metabolizing microbial cells.

Though metabolic pathways involve many reactions catalyzed by multiple enzymes, the single-enzyme kinetic formulation (Eq. (11.30)) is often appropriate, as the overall reaction is often limited by a single step. When a reactive pathway involves multiple electron donors and/or multiple electron acceptors, the respective limitation of the reaction rate is commonly described using a combination of the respective Michaelis–Menten terms. Together, they can be combined into a non-dimensional *kinetic factor* F_K (Fig. 11.7):

$$F_K = \prod_i \frac{S_i}{K_m^i + S_i} \tag{11.32}$$

The parameters K_m^i characterize the typical half-saturation concentrations for the associated substrates. Some other formulations are also available [5]. Many diagenetic models employ also *inhibition factors*, which may have similar mathematical shapes. For instance, the toxicity of oxygen to sulfate reducing bacteria could be described by a decrease in the rate of sulfate reduction as

$$R = V_{\max} \frac{[SO_4^{2-}]}{K_m + [SO_4^{2-}]} \frac{K'}{K' + [O_2]} \tag{11.33}$$

The inhibition constant K' here corresponds to the concentration of oxygen at which the rate of sulfate reduction decreases by a factor of two relative to the completely anoxic conditions ($[O_2] = 0$). In contrast to the Michaelis–Menten kinetic term, however, the functional form of the inhibition term is chosen for its mathematical convenience, without a firm mechanistic basis.

At very low concentrations of substrates, the power supplied by the catabolic reaction may fall below the minimum required to synthesize ATP for the microbial cell. Below that threshold, the reaction would become thermodynamically unfeasible for the microbe, even if the overall ΔG of the catabolic reaction is favorable. The corresponding slowdown in the reaction rate is described with a dimensionless *thermodynamic factor* F_T [198] (Fig. 11.7):

$$F_T = 1 - \exp\left(\frac{\Delta G_{\mathrm{cat}} + m\Delta G_{\mathrm{ATP}}}{\chi RT}\right) \tag{11.34}$$

The available catabolic energy ΔG_{cat} is adjusted in this equation by the energy that is conserved by the microbe in the form of ATP, for anabolic and maintenance needs. m is the number of mols of ATP synthesized per reaction and χ is a pathway-dependent stoichiometric factor.

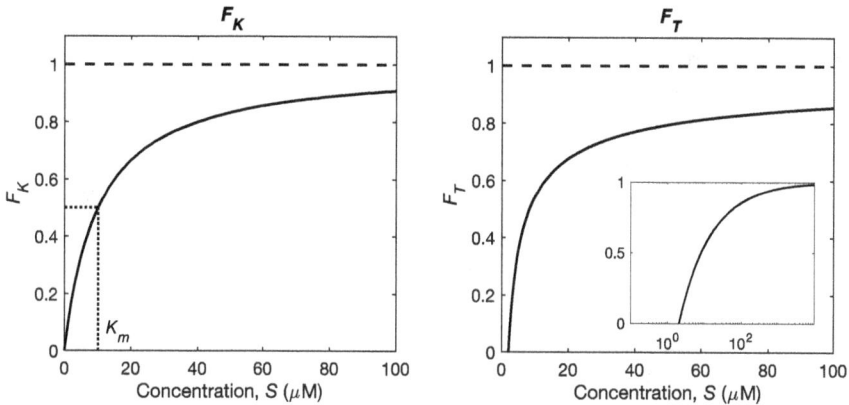

Fig. 11.7. Kinetic and thermodynamic factors for microbially-catalyzed reactions. The values used in the calculation were chosen for illustration, for catabolic energies similar to those of sulfate reducers: $K_m = 10~\mu$M, $\Delta G_{ATP} = 50$ kJ/mol, $m = 1$, $\chi = 2$, $T = 277$ K. The inset shows the same graph for F_T with substrate concentration plotted on a logarithmic scale.

Amending the Monod equation using the kinetic and thermodynamic factors above, the net rate of a microbially-catalyzed reaction can be written as

$$R = V_{\max} F_K F_T \qquad (11.35)$$

The reaction rate constant $V_{\max}$ reflects the kinetic capabilities of the microbial cells and the abundances of cells present in the environment. As the total concentration of an enzyme E generally scales with the size of the microbial population, the rate can be written as

$$R = rX = v_{\max} F_K F_T X \qquad (11.36)$$

where r is the *cell-specific* rate (mol$_{rx}$ cell^{-1}h^{-1}) of the reaction. Here X is the biomass concentration (e.g., the number of cells per unit volume) of the relevant microbial population and $v_{\max}$ is the cell-specific rate (i.e., the reaction rate per unit of biomass). Using $v_{\max}$ rather than $V_{\max}$ may be advantageous, as it may relate to the specific physiological capabilities of a given type of organism, which could be potentially transferable between environments. In contrast, $V_{\max}$ depends on the microbial biomass and may vary in a wider range.

In dynamic environments, the biomass X can change with time. Microbes grow when conditions become favorable or die under adverse conditions. When conditions change, reaction rates may respond with a delay, as new microbial populations need time to grow. The rate of change for the biomass can be described as the difference between the growth and death rates in the population:

$$\frac{dX}{dt} = rYX - \lambda X \qquad (11.37)$$

where Y is the yield (mol_X/mol_{rx}) (see Chapter 4), i.e., the number of mols of biomass synthesized per mol of the catabolic reaction. The parameter λ (1/h) describes the decay rate for the living biomass. Note that the bulk rate of the geochemical reaction, $R = rX$, depends on the size of the population (Eq. (11.36)), while the population biomass X, in turn, grows by carrying out the reaction. Microbial reactions are thus, in essence, autocatalytic [5]. The death parameter λ, in principle, may include the dependence of mortality on a variety of environmental parameters.

Table 11.1 lists some typical values for the kinetic parameters that have been obtained for main diagenetic pathways in different studies.

11.3.3 Abiotic reactions

Reactions that do not need to explicitly involve the dynamics or physiology of microbes can be typically described by one of the simpler reaction laws, such as the first- or second-order kinetics (Section 3.6.2). This may be appropriate also for the microbially mediated reactions the rates of which are not strongly affected by physiological limitations, at the concentrations of interest (e.g., Table D.2). For instance, reactions that involve oxidation of the dissolved products of anaerobic metabolisms can be often described as second-order reactions with respect to the concentrations of the oxidant and the reductant, e.g.,:

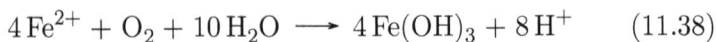

$$4\,Fe^{2+} + O_2 + 10\,H_2O \longrightarrow 4\,Fe(OH)_3 + 8\,H^+ \qquad (11.38)$$

$$R_{FeOx} = k_{FeOx}[Fe^{2+}][O_2] \qquad (11.39)$$

For a known concentration range of the reactants, the characteristic time scale on which the reaction occurs can be evaluated from the value of the reaction rate constant. For instance, let's say that the

Table 11.1. Selected microbial kinetic parameters. Values in brackets are those reported in cultures, rather than from natural environments. The subscript "X" denotes microbial biomass.

Parameter	Value	Unit	Ref.
v_{max}			
Denitrifiers	0.2	$mol_{rx}/mol_X/h$	[199]
Iron reducers	0.01	$mol_{rx}/mol_X/h$	[27]
Sulfate reducers	1.0	$mol_{rx}/mol_X/h$	[200]
Methanogens	1.0	$mol_{rx}/mol_X/h$	[200]
K_m for e^- acceptor			
Aerobes (O_2)	2	μM	
Denitrifiers (NO_3^-)	10	μM	[199]
Manganese reducers (MnO_2)	4–32	$\mu mol/g_{dw}$	[201]
Iron reducers (FeOOH)	1000^1	$\mu mol/L$	[27]
Sulfate reducers (SO_4^{2-})	5; 200	μM	[166]
K_m for acetate			
Denitrifiers	10	μM	[199]
Iron reducers	0.8	μM	[202]
Sulfate reducers	0.1–5 (70–230)	μM	[203–205]
Methanogens	12–50 (400–700)	μM	[202, 204, 205]
K_m for hydrogen			
Sulfate reducers	(1–4)	μM	[204]
Methanogens	(5–7)	μM	[204]
K'_{O2}	2	μM	

vertical profile of Fe^{2+} extends upward from the reduced sediment and overlaps with the lower part of the oxic layer in a region where the concentrations of both Fe^{2+} and O_2 are at low micromolar levels. The typical value of the reaction rate constant k_{FeOx} on the order of 100 $\mu M^{-1} y^{-1}$ (Table D.2) then implies that the oxidation reaction may take several days. This is slow enough to allow some of the dissolved iron to diffuse through that region into the overlying sediment layers. Higher (100s of μM) concentrations of oxygen near the sediment surface, however, would cut the oxidation time by an order of magnitude, down to several hours or shorter, effectively preventing iron from reaching the overlying water column.

---

[1] Depends on Fe mineralogy; may be smaller for nanophase precipitates.

Experimentally determined values of the reaction rate constants are scattered throughout geochemical literature. While some attempts have been made to compile such values in databases [206], a common process for obtaining rate estimates for diagenetic processes is still to comb the literature for process-specific parameters. Besides experimental studies, modeling papers often include conveniently arranged tables of reaction rate constants. Values for the reaction rate constants may be compiled from experimental studies (performed either in sediments *in situ* or in carefully prepared systems in a laboratory), or they may be obtained in modeling studies as model's fitting parameters. Comparing the values from several sources usually allows one to constrain their ranges within at least an order of magnitude. Some values for the reaction rate constants compiled from literature are given in Table D.2.

11.3.4 Mineral precipitation and dissolution

Precipitation of minerals, such as FeS or vivianite, is sometimes considered similarly to the abiotic reactions above, e.g. as second-order reactions with rates directly proportional to the concentrations of the participating chemical species. There are, however, important differences, as the rate of precipitation necessarily slows down when the mineral is close to being in equilibrium with the solution. When concentrations in the solution are so low that the solution is undersaturated with respect to the mineral, dissolution takes place instead of precipitation. The rate of the dissolution may follow a different kinetic law than that for precipitation.

According to the theory described in Section 3.6.3, precipitation rate is proportional to the degree of supersaturation. At high concentrations of solutes, the rate is more or less proportional to the concentrations (or activities). For instance, precipitation of FeS may be a second-order reaction with the rate proportional to the concentrations of Fe^{2+} and HS^-. (As the reaction of FeS precipitation involves releasing a proton H^+ and can be written with either HS^- or H_2S, the kinetics would also depend on the pH [207].) When one of the components of a binary mineral is in abundance, the rate may be considered as being first-order. For example, when carbonate ions are abundant, precipitation of calcium carbonate may be considered as a first-order process controlled by the Ca^{2+} concentration (and the pH). At low concentrations, however, the solubility equilibrium needs to be taken into account. A viable description, for example, considers the precipitation rate to be proportional to the degree of supersaturation:

$$R_{prec} = k_{prec} \left(\frac{\Omega}{K_{eq}} - 1 \right) \tag{11.40}$$

where Ω is the activity product of ions in the solution, which must exceed the equilibrium *solubility product* K_{eq}. The reaction rate constant k_{prec} in this case would have the dimension of rate (e.g., mol per liter per year).

An analogous formulation for the rate of dissolution would be

$$R_{diss} = k_{diss} [C_{min}] \left(1 - \frac{\Omega}{K_{eq}} \right) \tag{11.41}$$

where $[C_{min}]$ is the concentration of the mineral and Ω is smaller than K_{eq}. (Note that the dimension of the reaction constant in this case is 1/time.) In contrast to the rate of precipitation, however, the dissolution rate may not be necessarily a strong function of undersaturation (Section 3.6.3). The dissolution rate then may be considered as simply proportional to the amount of the mineral, as a first-order kinetics.

11.3.5 Adsorption

The reactions of sorption and desorption are often fast and can be considered to be in local equilibrium. The adsorbed amounts can be described in several ways.

A simple descriptive way is to specify the *distribution coefficient* (Table 11.2), K_d, defined as the ratio of the concentrations of the sorbed and dissolved fractions:

$$C_{ads} = K_d C_{sol} \tag{11.42}$$

When the concentrations for the adsorbed and aqueous species are measured in different units (e.g., mol per mg of substrate vs. mol per ml of porewater), the distribution coefficient can be made into a dimensionless quantity by explicitly incorporating the sediment porosity ϕ and solid sediment density ρ [208, 209]. These linear adsorption coefficients, while relatively easily measured, cannot

Table 11.2. Linear adsorption coefficients reported in marine clay sediments [1].

Species	K_d	Ref.
Ca^{2+}	1.4–1.8	[210]
NH_4^+	1.3–1.6	[86]
PO_4^{3-}	1.2–2.5	[209]

account for the dependencies of the adsorption capacity on aqueous concentrations, pH, or other factors. They are unlikely to accurately describe the adsorption of ionic species.

A slightly more refined, but still purely empirical, description is given by the *Freundlich isotherm*:

$$C_{\text{ads}} = K_f C_{\text{sol}}^{n_f} \qquad (11.43)$$

The exponent $0 < n_f < 1$ accounts for decreasing sorption at higher solute concentrations, as most of the readily available sorption sites on mineral surfaces become occupied.

The *Langmuir isotherm* accounts for a finite number of sorption sites, as well as for the competition among aqueous species for those sites. The adsorbed fraction is then described as

$$C_{\text{ads}} = C_{\text{max}} \frac{k_l C_{\text{sol}}}{1 + k_l C_{\text{sol}}} \qquad (11.44)$$

where C_{max} is the maximum adsorption capacity and the parameter k_l describes the surface's affinity[2] towards the substance. At small k_l or low concentrations when $k_l C_{\text{sol}} \ll 1$, the Langmuir isotherm reduces to a *linear isotherm* analogous to Eq. (11.42) [1].

11.3.6 Bioturbation

Bioturbation is commonly represented as particle diffusion. The corresponding diffusion coefficient D_b is often a function of depth into

---

[2]In equilibrium, it may be linked alternatively to the kinetic ratio of the sorption and desorption rates, or, equivalently, to the thermodynamic binding energy of the adsorbed complex.

the sediment, as most activity occurs near the sediment surface. Approximating bioturbation as diffusion, however, is not an obvious choice. Benthic organisms displace solid sediment particles in a number of ways. Burrowing fauna move them around by small distances. Worms ingest them, past them through their gut and excrete some distance away. Other organisms take particles from the reduced sediment and deposit them as a mound on the sediment surface. If one were to trace the position of a single mineral grain within the bioturbation zone, one would observe that it spends most of its time at rest, while occasionally being moved by irregular distances. Nevertheless, on time scales longer than about several months [211], the net result of all these processes is well represented by a "random walk", similar to diffusive Brownian motion.

The diffusion coefficient is generally observed to correlate with the sedimentation flux of organic carbon, which ultimately fuels the benthos activity. The relationship, however, is rather noisy and can be used only for rough predictions. Only the upper few cm of freshwater sediments are bioturbated, whereas marine sediments can be bioturbated much deeper, by 20 cm and more. The values of D_b near the sediment surface, however, appear to be similar between lakes and marine environments.

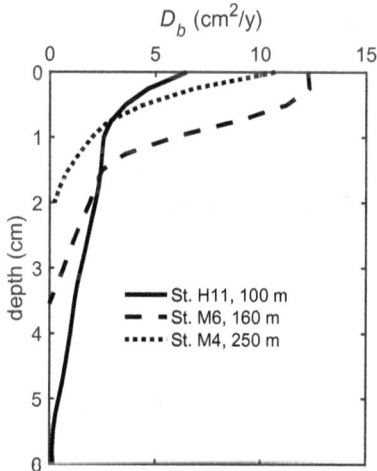

Fig. 11.8. The intensity of bioturbation $D_b(z)$ at several locations in lakes Michigan and Huron, calculated from the vertical profiles of oxygen and organic carbon using Eq. (11.47).

Example: Estimating bioturbation intensity. Rates of bioturbation are notoriously difficult to quantify and are commonly characterized using experimental approaches, such as sediment incubations with particle tracers [17]. The diagenetic equation, however, provides a way of estimating the value of the bioturbation coefficient D_b in select conditions, when sufficient amounts of geochemical data are available (Fig. 11.8) [11]. In particular, Eq. (11.10) can be used, as it describes the effects of bioturbation on the distribution of organic carbon. When a steady state can be assumed, at some depth L deep enough within the sediment the concentration C of organic carbon should become constant $((dC/dz)_L = 0)$, and bioturbation also ceases $(D_b(L) = 0)$. The equation then can be integrated over depth, from L to some given depth z:

$$\phi_s(z)D_b(z)\left(\frac{dC}{dz}\right)_z - [\phi_s(z)v(z)C(z) - \phi_s(L)v(L)C(L)]$$
$$- \int_L^z \phi_s(z')R_C(z')\,dz' = 0 \tag{11.45}$$

where R_C is the rate of organic carbon mineralization. The integral corresponds to the total rate of mineralization below the depth z. As discussed in Section 5.1.6, it can be approximated by the downward flux of oxygen at that depth:

$$F_{O_2}(z) = -\phi D_{O_2}\frac{d[O_2]}{dz} \tag{11.46}$$

where D_{O2} is the molecular diffusion coefficient for oxygen (this neglects bioirrigation). The bioturbation coefficient is then

$$D_b(z) = \frac{[\phi_s(z)v(z)C(z) - \phi_s(L)v(L)C(L)] - \phi D_{O_2}\left(\frac{d[O_2]}{dz}\right)_z}{\phi_s(z)\left(\frac{dC}{dz}\right)_z} \tag{11.47}$$

At a steady sedimentation rate, the expression can be simplified further, as $\phi_s(L)v(L) = \phi_s(z)v(z)$ (Eq. (2.5)). Thus, having high-resolution profiles of oxygen concentrations, porosity, the sedimentation rate, and the concentrations of organic carbon in the sediment allows one to calculate $D_b(z)$—the intensity of sediment bioturbation and its variation with depth in the sediment.

11.3.7   Bioirrigation

The non-local term R_{irr} in Eq. (11.7) most commonly refers to bioirrigation. A commonly used approximation, for example, is

$$R_{\text{irr}}(z) = \alpha_{\text{irr}}(z)(C_0 - C_{\text{burr}}) \qquad (11.48)$$

where C_0 is the solute concentration immediately above the sediment and C_{burr} is the concentration inside animal burrows. For practical purposes, it is often taken as $C_{\text{burr}} = C$. The function $\alpha_{\text{irr}}(z)$, which typical decreases with depth, is called the *coefficient of bioirrigation*. This formulation, sometimes called "the alpha model" [212, 213], is widely used, though it may become inadequate when addressing complex reaction networks, such as for the coupled dynamics of oxygen and sulfide. More complex (e.g., two- or three-dimensional; Figs. 2.5, 2.6) descriptions that account for the structure of animal burrows may be require in those cases.

11.4   Numerical Models

The theoretical foundations for describing sediment diagenesis mathematically were laid out by Robert Berner in the 1980s [1]. With advances in computational power, pioneering numerical models appeared in the 1990s [208, 214]. The models solved systems of diagenetic equations (Eqs. (11.7)–(11.8)) for selected sets of chemical species. As the equations couple geochemical reactions and physical transport, models of this type became known as *reaction-transport* (or "reactive transport") models (RTMs). To parameterize the rates of the simulated reactions, where possible, these vertically-explicit models used available experimental data, while some of the less-known parameters were being treated as fitting parameters. Collections of parameters from different studies were occasionally compiled in databases [206].

A typical application of an RTM involves calibrating the model to reproduce the observed vertical distributions of chemical species (Fig. 11.9). The corresponding geochemical rates and fluxes then may be obtained directly from the model. Being explicit components of the diagenetic equation (Eqs. (11.7)–(11.8)), the rates R_i of the individual model reactions can be obtained as functions of depth z within

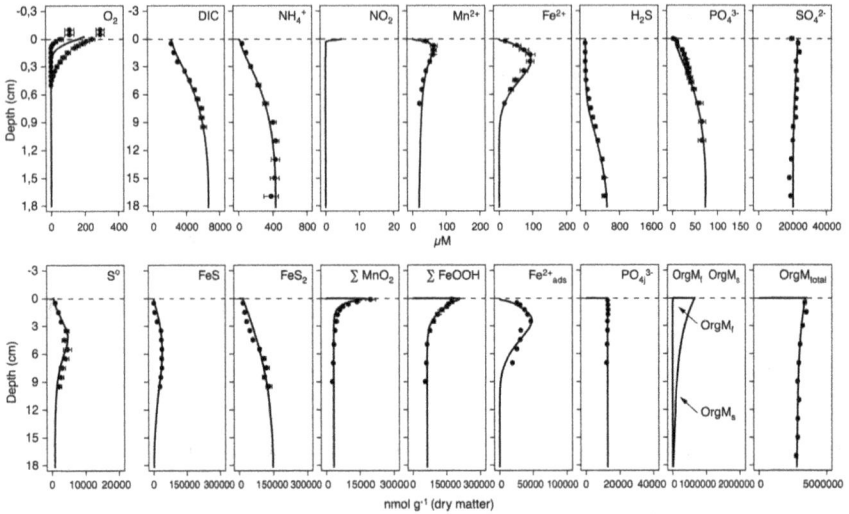

Fig. 11.9. An example of output from a numerical reaction-transport model (for a marine sediment in Aarhus Bay, Denmark; from [217]) Symbols are measured data and lines are model solutions.

the sediment (and time t, if not at steady state). Fluxes can be similarly calculated from the modeled concentration gradients and/or volumetric rates (Section 11.2.1). Importantly, models help understand processes that are not directly accessible through experimental techniques, such as the rates of highly coupled reactions. They can also be used to project the future dynamics of the system or its responses to perturbations. Beyond site-specific studies, models can be used in an exploratory fashion [136], to test geochemical theories and evaluate the sensitivity of geochemical systems to environmental conditions. In such explorations, model parameters are varied to investigate geochemical outcomes over a range of conditions, including in aquatic environments for which only minimal or no data exist [215] [32].

By their nature, RTMs consider multiple reactions with multiple kinetic parameters, which means that they are typically *overparameterized* [216]: certain fitting parameters cannot be fully determined from the available calibration data. Simulation results may not always be unique, with similar solutions being obtained for different sets of parameters, each within the realm of possibility. To identify more general, robust trends, *sensitivity analyses* are performed, whereby model solutions are investigated over a range of model's parameters.

Recent advances in modeling techniques (e.g., reviewed in [218]) saw the development of user interfaces, publicly available codes, and model implementations in multiple programming languages. Building on the success of earlier geochemical models, *biomass-explicit* models [30, 219] incorporated functional microbial populations as model variables, to simulate their biomasses and metabolic activities. With exponential advances in the availability of genomic and multi-omic datasets, a further cohort of models has started to incorporate formulations that explicitly account for microbial activities, as measured by the abundances or expression rates for the corresponding functional genes [44, 220]. As gene expression rates reflect the rates of cell multiplication, which in turn are tied to the geochemical rates of microbial catabolisms, such *genome-enabled* models can use the available omics information for model calibration, and can use geochemical information to gain insights into microbial metabolisms. In another development, models that explicitly consider multiple isotopes (e.g., of S or N), are being applied in the interpretations of the isotopic signals preserved in sediments [221].

A potentially important but rarely explored direction in diagenetic modeling is the dynamic coupling between sediments and the overlying water column [222,223]. Feedbacks between processes in the two domains may strongly influence their coupled dynamics, resulting in hysteresis or multistability in the system's dynamic behavior [136, 224]. For example, eutrophic (high biological productivity) conditions in the water column may lead to excessive consumption of oxygen near the bottom, which facilitates the mobilization of phosphorus from sediments. The higher availability of phosphorus, in turn, fuels biological productivity, exacerbating the anoxia. Lakes subjected to such conditions may be more difficult to restore to a low-productivity regime, which is often more desired from an environmental management perspective.

Chapter 12

Solutions to Exercises

12.1   In-Chapter Examples

Chapter 2

- Page 9: For an approximate answer, we may take the average porosity of about 0.9. The volume of our sediment sample is $V_{sed} = 10 \times 100 = 1000\,\text{cm}^3$, and the corresponding volume of the porewater is $0.9 \times 1000 = 900\,\text{cm}^3 = 900\,\text{mL}$. For a more accurate result, one would need to sum the volumes of the porewater in each horizontal layer of sediment.
- Page 11: Assuming a typical solid sediment dry density $\rho = 2.6\,\text{g/cm}^3_{dw}$, the mass of the solid material in the top 0.5 cm slice is $\rho(1 - \phi) = 2.6 \times (1 - 0.96) = 0.104\,\text{g/cm}^2$. It will take $0.104\ \text{g cm}^{-2}/0.014\ \text{g cm}^{-2}\text{y}^{-1} = 7.42$ years to accumulate this much sediment, corresponding to the average burial velocity of $0.5/7.42 = 0.067$ cm/y (close to the value shown in Fig. 2.3). At the depth of 4 cm, the volume fraction of solid sediment is significantly greater, approximately $\phi_s = 0.1$. The corresponding thickness is then just $0.5 \times 0.04/0.1 = 0.2\,\text{cm}$, and the burial velocity is only $0.067 \times 0.04/0.1 = 0.027\,\text{cm/y}$.
- Page 15: The time scale of diffusion (at $T = 4°\text{C}$) is $(2\,\text{cm})^2/(2 \times 371\,\text{cm}^2/\text{y}) = 0.00539\,\text{y} \approx 47\,\text{h}$. Note that this only gives an approximate (characteristic) time for the diffusion of bromide molecules, and traces of bromide may be detected at that distance earlier.

Chapter 3

- Page 27: The flux of ammonium out of the layer balances the production of ammonium within the layer. The production rate per volume is 10 μmol/L/h $= 10^{-9}$ mol/cm^3/h. As the layer is 1 cm thick, the flux through each square centimeter of the cross-section is 10^{-9} mol/cm^2/h.
- Page 32: From the Gibbs free energy of formation for the reactants and products (Appendix C), the energy of the reaction under standard conditions is

$$\Delta G^0 = -744.8 - 27.89 - 2 \times 16.32 - 2 \times (-237.18) = -330.97 \text{ kJ/mol}$$

The reaction quotient at equilibrium is

$$Q = \frac{[SO_4^{2-}][H^+]^2}{[H_2S][O_2]^2} = e^{-\frac{\Delta G^0}{RT}}$$

At pH 7 and $T = 4°C = 277$ K,

$$\frac{10^{-4}(10^{-7})^2}{[H_2S](10^{-6})^2} = e^{-\frac{-330.97}{8.31 \times 10^{-3}\,277}}$$

which gives $[H_2S] = 3.6 \times 10^{-69}$ M. So, at equilibrium, virtually no hydrogen sulfide can coexist even with trace amounts of oxygen.

- Page 35: The equilibrium between the gas and aqueous phases is described by Henry's law (Section C.4). The concentration of aqueous CO_2 is

$$[CO_2(aq)] = (3.4 \times 10^{-2}) \cdot (0.412 \times 10^{-3}) = 1.4 \times 10^{-5} \text{ M}$$

So, the equilibrium concentration (at 25°C) is 14 μM. The aqueous CO_2 makes only a minority of the total dissolved pool, however. The total dissolved CO_2 can be calculated from Eq. (C.1). It yields 76.6 μM at pH 7 and 643 μM at pH 8. At T= 4°C, these numbers are 141 and 1183 μM, respectively. These concentrations are lower than typical concentrations of DIC in the surface waters of lakes and oceans (a few mM), suggesting that the CO_2 could be diffusing into the bubble. Calculating gas exchanges with the atmosphere, i.e., for an open system, requires additional corrections. Also, large uncertainties exist about the values of the equilibrium

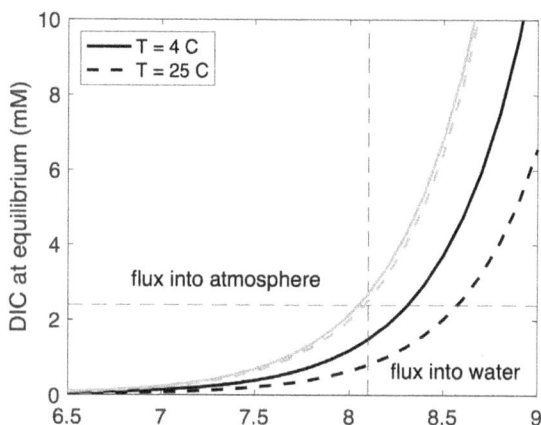

Fig. 12.1. Calculated concentrations of DIC in equilibrium with atmospheric CO_2. Reference values (black) vs values determined in the environment (grey) (from [225] for the ocean). The crosshair marks the values of the pH and DIC often quoted as typical for the surface ocean.

constants under environmental conditions, due to the effects of ionic strength, temporal and spatial variations in pH, temperature, biological activity, and other factors (Fig. 12.1). Accounting for such corrections is an active area of research, and accurate estimates can become quite complicated [225].

- Page 52: The time scale of diffusive transport over several cm is several days (Table 2.2). So oxygen is being transported to the reaction site much faster than it can be consumed by the reaction. The rate is therefore reaction-controlled, and the oxidation of organic carbon by oxygen would proceed at the rate dictated by the reactivity of organic carbon.

Chapter 4

- Page 65: Thermodynamic favorability of hydrogenotrophic metabolisms: iron reduction vs. methanogenesis. The corresponding reactions (Table D.1) and the Gibbs free energy gains under standard conditions (in kJ/mol) are as follows.

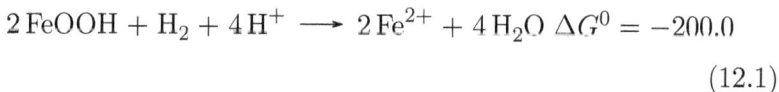

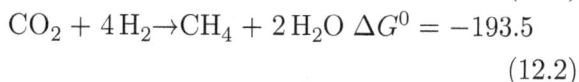

$$2\,FeOOH + H_2 + 4\,H^+ \longrightarrow 2\,Fe^{2+} + 4\,H_2O \ \Delta G^0 = -200.0$$

$$(12.1)$$

$$CO_2 + 4\,H_2 \rightarrow CH_4 + 2\,H_2O \ \Delta G^0 = -193.5$$

$$(12.2)$$

For ferrihydrite as the reactive Fe(III) phase, we obtain that the corresponding equilibrium conditions are given by

$$Q = \frac{[Fe^{2+}]^2}{[H_2][H^+]^4} = e^{-\Delta G^0/RT} \tag{12.3}$$

$$Q = \frac{[CH_4]}{[CO_2][H_2]^4} = e^{-\Delta G^0/RT} \tag{12.4}$$

At pH 7, T = 277 K, and for typical concentrations $[Fe^{2+}] = 10^{-5}, [CH_4] = 10^{-5}, [CO_2] = 10^{-5}$, this gives the threshold concentrations above which the reactions are favorable of

$$[H_2] = 10^{-19} \, M \tag{12.5}$$

$$[H_2] = 10^{-9} \, M \tag{12.6}$$

For less reactive iron oxides, the threshold concentrations of hydrogen are somewhat higher. The calculation shows, however, that hydrogenotrophic methanogens require at least nanomolar hydrogen concentrations, whereas iron reducers can extract energy using hydrogenotrophic metabolisms at lower hydrogen concentrations. This gives them ability to outcompete methanogens for hydrogen.

Chapter 8

- Page 119: Each cubic centimeter of sediment contains $(1 - 0.9) \times 2.65 \times 0.01 = 0.00265$ grams of organic carbon, or 220.8 μmolC. For the Redfield C:P ratio in marine phytoplankton, this corresponds to $220.8/106 = 2.08$ μmol of phosphorus, mineralization of which at 50% efficiency would yield 1.04 μmolP per 0.9 cubic cm of porewater. This corresponds to the phosphate concentration of 1.16 mM. Obviously, diffusion and mineral precipitation prevent such high concentrations in real sediments.

Chapter 11

- Page 165: As the sediment is at steady state, the fluxes at the top and bottom of the diagenetically active zone must balance. Only 10% of the deposited P is being buried, which means that the other 90% must be removed back to the water column: the vertical flux must be 0.09 mmolP m^{-2} y^{-1}. Most of this flux may be assumed to be phosphate, though some P may be contained in DOM. No additional knowledge about processes within the sediment is needed, as long as the steady state holds.

- Page 167: At steady state, as the methane produced within the methanogenic zone diffuses upward, its depth-integrated rate of production ($\text{mmol m}^{-2}\text{y}^{-1}$) should be equal to its flux towards the oxidation zone. The flux can be calculated using Fick's law of diffusion (Eq. (2.10)) from the approximately linear concentration gradient. Using the diffusion coefficient of 490 cm^2y^{-1} (Table 2.1; if higher accuracy is needed, a tortuosity correction may be done using Eq. (2.11)), this yields

$$F = 420 \times \frac{(0.5 - 0) \times 10^{-3}}{50} = 4.2 \times 10^{-3} \, \text{mmol cm}^{-2}\,\text{y}^{-1} \quad (12.7)$$

or 42 $\text{mmol m}^{-2}\text{y}^{-1}$. The factor 10^{-3} accounts for the conversion of concentrations from mmol/L to mmol/cm^3.

12.2   Selected End-of-Chapter Exercises

Chapter 2
4. Consider the typical shape of the oxygen profile in the sediment Fig 12.2. Noting that the depth of oxygen penetration is rather small in this case, we may expect that the oxygen profile is close to a straight line. The straight dashed line in the diagram therefore may provide a reasonable approximation for the gradient in the oxygen concentration near the sediment-water interface. The actual gradient is expected to be somewhat steeper, so the gradient obtained from this line is going to be a conservative minimum estimate:

$$\frac{d[O_2]}{dz}\Big|_{z=0} = \frac{0 - 100}{0.2 - 0} = -500 \, \text{nmol cm}^{-4}$$

This gradient determines the diffusive flux of oxygen into the sediment:

$$F_{O2} - -\phi D_{O_2}\frac{d[O_2]}{dz}$$

$$= 0.9 \times 671 \times 500 = 302{,}000 \, \text{nmol cm}^{-2}\text{y}^{-1} = 8.3 \, \text{mmol m}^{-2}\,\text{d}^{-1}$$

Here, we used the value for the molecular diffusion coefficient of oxygen at 4°C from Table 2.1 and a typical porosity ϕ of 0.9. Again,

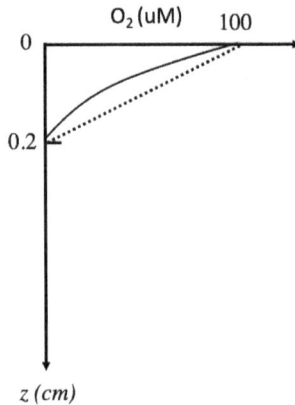

Fig. 12.2. Diagram for Problem 4 in Chapter 2.

this number is a conservative minimum estimate, as non-diffusive processes such as bioirrigation may also contribute to the oxygen flux.

5. We will assume, for simplicity, that the 20% drop in oxygen concentration is small enough so that the flux of oxygen into the sediment remains approximately constant. The total amount of oxygen in the overlying water is $(200\,\mu\text{mol/L}) \times (1L) = 200\,\mu\text{mol}$. At the area-specific consumption rate of 20 mmolO$_2$ m^{-2} d^{-1}, given the surface area of $\pi d^2/4 = 78.54$ cm^2, the consumption rate is $20 \times 78.54 \cdot 10^{-4}\,\text{m}^2 = 0.1571$ mmolO$_2$ d^{-1}. The time for the oxygen concentration to drop by $0.2 \times 200 = 40\,\mu\text{mol/L}$ is then $40/157.1 = 0.25$ d ≈ 6 hours.

Chapter 5

3. The mass balance at steady state requires that the carbon that is deposited to the sediment surface is either buried into the deep sediment or mineralized and released as CO$_2$. The burial efficiency, by definition, is the ratio of the burial and depositional fluxes. The total oxygen uptake (TOU) approximately corresponds to the depth-integrated rate of C mineralization. The burial flux of organic carbon can be calculated (Eqs. (2.6) or (2.7)) as

$$F_{\text{bur}} = v_z C = 1\frac{\text{mm}}{\text{y}} \times 0.02\frac{\text{gC}}{\text{gdw}} \times \left(2.6\frac{\text{gdw}}{\text{cm}_{\text{dw}}^3} \times (1-0.8)\frac{\text{cm}_{\text{dw}}^3}{\text{cm}^3} \right)$$

Here, we assumed the density of dry sediment of 2.6 g/cm^3. Expressed in the same units as the oxygen uptake (with the molar mass of carbon being 12 $\frac{gC}{molC}$, this flux amounts to approximately 2.4 mmol/m^2/d. From mass balance, the total rain flux of carbon should be $2.4 + 5.0 = 7.4$ mmol/m^2/d. And the burial efficiency is therefore $2.4/7.4 = 0.32 = 32\%$.

4. See the solution to Exercise 4 in Chapter 2. The flux of oxygen into the sediment is approximately stoichiometrically equivalent to the depth-integrated rate of mineralization of organic carbon.

Chapter 6

5. Nitrification is the oxidation of ammonium by oxygen in the oxic zone. While some ammonium is being produced by mineralization of organic matter within the oxic zone, in this case the oxic layer only spans a few mm, so for an approximate calculation it may be appropriate to approximate the nitrification rate as the oxidation of ammonium that diffuses upward from the deeper, anoxic layers. Using the diffusion coefficient for ammonium of 352 cm^2 y^{-1} (Table 2.1) and the concentration gradient near the oxidation zone boundary, estimated from the figure at around 50 μM/cm, gives the upward diffusive flux of ammonium (Eq. (2.10)) of around 0.50 mmol/m^2/day. This flux could be then taken as a measure of the total, depth-integrated, rate of sediment nitrification (in mmol N/m^2/day). To estimate volumetric rates, one would need to divide this number by the estimated width of the nitrification zone, likely a few mm in this case. Likewise, assuming that denitrification immediately below the oxic-anoxic boundary is the dominant process that removes the downward-diffusing nitrate, the depth-integrated rate of denitrification may be estimated from the concentration gradient of nitrate near the oxic-anoxic boundary, around 1 cm depth. This estimate produces the depth-integrated rate of denitrification of around 0.16 mmol N/m^2/day. For comparison, the measured rates of denitrification at that site reported in Ref. [77] were around 0.20 mmol N/m^2/day. One may also place this result in the context of Fig. 6.5, as the measured oxygen uptake reported in that study was around 5 mmol O$_2$/m^2/day.

Chapter 7

2. The redox potentials plotted in Fig. 3.4 indicate that the reduction of Fe oxides is not possible with ammonium. However, the energetic feasibility of Fe-ammox needs to be verified for a range of environmental concentrations and for multiple alternative combinations of

redox pairs, i.e., for different mineral phases of Fe(III) and different oxidized forms of N (Table C.6).

3. Conditions for siderite precipitation can be calculated from the corresponding solubility product (Table C.3). For pure siderite (at standard temperature and pressure):

$$[Fe^{2+}][CO_3^{2-}] = 10^{-10.45} \qquad (12.8)$$

For the concentrations of ferrous iron on the order of 100 μM = 10^{-4}M, supersaturation requires carbonate ion activities on the order of $10^{-6.45}$, i.e., 0.3 μM. At pH 7, this corresponds (Eq. (C.3)) to 12 mM of DIC, which is too high for most environments. At pH 8, however, this requires only 1.4 mM of DIC, which is well within the typical range. Given than sediment porewaters are typically more acidic than overlying water column, this could suggest that siderite precipitation could be expected to be more common in marine environments and alkaline lakes than in lakes with neutral or acidic pH.

Chapter 8
4. The sediment's Fe-rich layer contains

$$0.01\frac{g_{Fe}}{g_{dw}} \times 2.5(1-0.9)\frac{g_{dw}}{cm^3} \times 0.5\,cm = 0.25\frac{mgFe}{cm^2}$$

or 45 mmol Fe/m^2. For a typical Fe:P molar ratio at the adsorption limit of, say, 10:1, reductive dissolution of this quantity of Fe(III) would correspond to the mobilization of 45 mmol P/m^2, over a period of 1 month. Most of this mobilized P could be expected to reach the water column, as any downward diffusion would likely be impeded by the typically high concentrations of phosphate in the reduced sediment. The P flux into the water column would thus be about $45/30 = 1.5$ mmol P/m^2/day. This is obviously much greater than any typical P fluxes under a steady-state situation: fluxes on the order of several mmol/m^2/day are more common for the exchanges of carbon (Chapter 5), and typical C:P ratios would suggest fluxes of phosphorus that are two orders of magnitude lower.

Chapter 10
1. Given the half-life of ^{210}Pb, halving of the decay activity occurs every 22.3 years. The burial velocity is thus $v = 3/22.3 = 0.13$ cm/y.

2. A porosity change from 0.98 to 0.96 corresponds to a two-fold increase in the amount of solid fraction per unit volume: from 0.02 to 0.04. Thus, half of the decay activity at the bottom of the 3-cm interval is generated by twice the amount of ^{210}Pb. The width of the interval thus corresponds to two half-lives, or 44.6 years, and the burial velocity is half of the value from the previous exercise.

Appendix A

Field and Laboratory Methods

A.1 Sediment Sampling

When sediment sampling is conducted in shallow environments, direct sampling may be possible by a diver pushing a sampling tube into the bottom. In very shallow environments, sampling may be done by a person standing in water, on ice, or from a boat or raft. In most cases, however, a specialized sampling device is needed (Fig. A.1). Most of the devices listed below are designed to resolve the one-dimensional vertical structure of sediment. The exceptions are the grab sampler, which yields a bulk (essentially, zero-dimensional) sample, and the box corer, which yields a sample large enough to preserve three-dimensional structures.

Grab sampler. The sampler scoops the upper few centimeters of sediment. Spatial structures and porewater are not well preserved by this method. The advantages are its ease of use and the substantial volume of collected sediment, as well as the relatively large surface area of sediment that gets sampled, which may be important when sampling for benthic animals.

Gravity corer. The heavy sampling tube is pushed into the sediment by its own weight. The corer can penetrate by up to 2 meters and preserves the vertical structure of the sediment relatively well, except in the upper centimeters which become disrupted by a rather rapid contact of the sampler with the sediment surface.

MUCK corer. The coring device, weighing several tens of kilograms, is attached to the top of the sampling tube. The cover is easily portable and can be used from small boats, and even deployed by hand without a winch. The disadvantage is that occasional jerks on the rope may trigger the corer while it is still in the water column, while waves rocking the boat may cause the device to land on its side. This limits its use in deep water.

Multicorer. This is the method of choice when sediment cores with a well preserved sediment-water interface are needed. The corer takes several simultaneously cores, penetrating to about 40 cm. The feet of the lander land first, after which the tubes are pushed into the sediment by the weight of lead bricks. The large size and weight of the device limit its use to relatively large boats.

Piston corer. The corer is capable of taking sediment cores up to 10 m in length, which makes it an important tool for paleoceanographic and paleolimnological studies. The corer uses a large weight (several hundred kilograms) and often uses a gravity corer or a MUCK corer as a trigger for the piston.

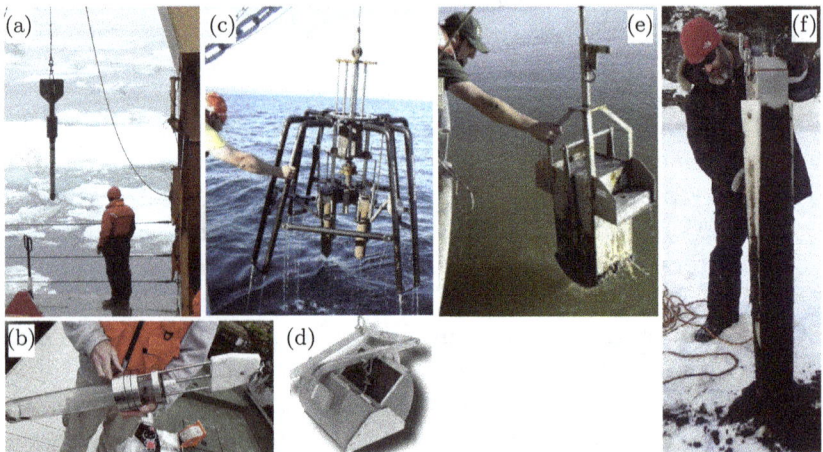

Fig. A.1. Sediment samplers: (a, b) Gravity corers; (c) Multicorer; (d) Ponar grab; (e) Box corer; (f) Freeze-corer. (*Image credits*: USGS (a, b, d, e) and this author (c, f).)

Push corer. The corer is used primarily in shallow environments. Multiple rod sections can be combined to extend its length, and the corer is pushed into the sediment by hand.

Box corer. The device can recover several hundred kilograms of sediment. The scoop at the lower end of the corer helps hold the sediment in place. Deployed from large ships, it can be used in studies of benthic fauna, or wherever large volumes of sediment are required.

Freeze corer. This corer uses a metal wedge filled with a powerful coolant (usually a mixture of dry ice (solid CO_2) and methanol). After the wedge is quickly lowered through the water column and penetrates the sediment under its own weight, it is left there for 20–40 min. The sediment freezes on the outside of the wedge, after which the corer is quickly recovered.

A.2 *In-situ* Methods

In addition to methods that recover sediment from the bottom for later analyses in the lab, several methods have been developed for investigating sediment properties *in situ* (Fig. A.2).

Benthic lander. Benthic landers are instrumental platforms that can be deployed onto the surface of the sediment, from where they can use their instruments to sample or monitor sediment conditions. Instruments that may be deployed this way include automatic samplers, video cameras, temperature and current monitors, and others. Some advanced techniques, for example, use several high-frequency, high-precision temperature and oxygen sensors positioned above the sediment surface to infer the oxygen fluxes into the sediment using the so-called eddy correlation technique [212].

Benthic chamber. This is a technique for measuring chemical fluxes out of the sediment. A portion of the sediment surface is covered with a chamber, and fluxes are determined from changes in chemical concentrations within the chamber over time. The drawback is that the chamber disrupts the hydrodynamic flow above the sediment, potentially impacting the fluxes.

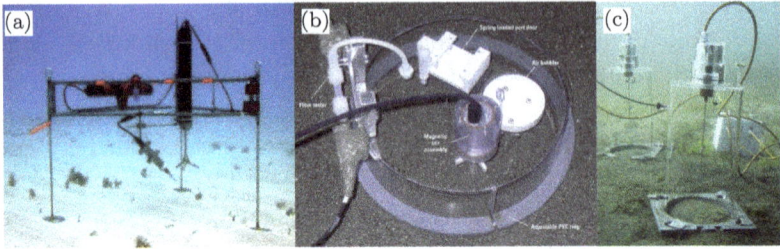

Fig. A.2. *In-situ* measurements and sampling: (a) Benthic lander with an eddy correlation system (*image credit*: Markus Huettel). (b, c) Benthic chambers (*image credits*: USGS [226] and Michael Bartz [227]).

A.3 Subsampling and Imaging Methods

A variety of methods exists for extracting porewater or solid phases for later analyses, and also for obtaining their characteristics in intact sediment samples (Fig. A.3).

Rhizon porewater extractor. Rhizon "artificial root" samplers have received widespread recognition as an efficient way to extract porewater from sediments [228]. Originally developed for extracting fluids from soils (hence the name), they use vacuum to draw porewater into an evacuated sampling vial or a syringe through a 0.1 μm filtering membrane. Rhizons can be used to sample directly from a sediment core by inserting the membrane-covered tip through a hole in the core wall. Alternatively, after a sediment core is extruded and wet slices of sediment are collected in sample tubes, rhizons may be used to extract the interstitial fluids from the tubes.

Peepers. Peepers sample the composition of sediment porewater *in situ*. An array of fluid-filled cells is buried into the sediment and left there for some time. The cells, covered by a permeable membrane, then equilibrate with the sediment porewater, so the concentrations of solutes within the cells approach those in the corresponding depths within the sediment. The peeper array is then recovered and the content of the cells is analyzed in the lab.

Microsensors. Whereas rhizons and peepers sample the sediment porewater with a spatial resolution of at most a few

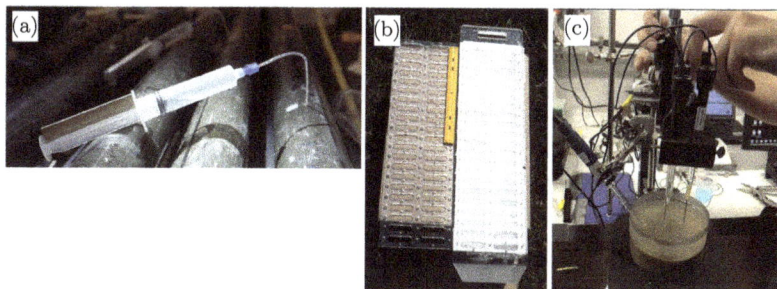

Fig. A.3. Subsampling and profiling: (a) Rhizon porewater extractors. (b) Peepers. (c) Microprofiling sensors. (*Image credits*: this author (a, c) and public domain (b).)

millimeters, microsensors afford the vertical resolution as fine as 0.1 mm or even smaller. The sensor is typically an electrode with a glass or needle-cased tip [229]. It is pushed into the sediment using a fine-calibrated micromanipulator that allows precise depth control. The vertical resolution of the resultant profile is defined by the diffusive field around the tip, which is about the size of the tip diameter, 50–500 μm. The technique can be used for obtaining vertical profiles of oxygen, hydrogen sulfide, the pH, redox potential (Eh), hydrogen (H_2), and other characteristics. Sensors have also been developed to detect concentrations of certain nitrogen species, such as N_2O and NO.

Planar optodes. When two-dimensional resolution is needed, some of the chemical distributions can be visualized—and even tracked in real time—by planar sensors. Most notably, planar optodes that can be placed along a wall of a sediment container, can track the concentrations of dissolved oxygen. This is especially useful in studies of sediment bioirrigation, where both the temporal dynamics and the three-dimensional structure are important [230]. Two-dimensional sensors sensitive to the concentrations of dissolved iron, phosphate, hydrogen sulfide, pH and several other sediment characteristics have also been developed [231, 232].

X-ray imaging. For three-dimensional imaging, sediment cores may be subjected to analyzes such as CT-scans, which help visualize internal structures, such as animal burrows [233].

A.4   Complementary Measurements in Water Column

Measurements in the water column above the sediment provide important complementary information about the factors that may affect sediment diagenesis (Fig. A.4). They include physical measurements of bottom currents, chemical measurements of oxygen concentrations, and a range of other parameters.

CTD+. Conductivity-temperature-depth (CTD) profilers are the workhorses of oceanography. They can be accompanied by a number of other profiling measurements, which obtain depth distributions of parameters such as dissolved oxygen, Chlorophyll, light intensity, fluorescence spectra (indicative of specific photosynthetic pigments), among others.

Water sampling. Water samples are recovered using containers that allow water to flow through them as they are lowered into the water column but can be closed at a specified depth by being remotely triggered. Several designs for such sampling bottles exist (e.g., Niskin, Van Dorn). On larger vessels, multiple bottles can be combined into a rosette sampler, often in combination with a CTD probe. A live signal cable allows operators to monitor the properties of the water column in real time and trigger the bottles electronically at a variety of chosen depths.

Current velocities. Bottom currents may be measured with instruments such as current meters or Doppler current profilers.

Remote imaging. Some important physical features of sediments may be resolved remotely from the ship. Sonars detect density gradients, and can help identify the thickness of the sediment pile above the bedrock, the density of the sediment surface or any characteristic layers within the sediment, or pockets of gas. Echo-sounding techniques in the water column, similar to commercially available fish finders, can help identify bubble plumes. Underwater cameras can be deployed to image the sediment surface from cabled instruments or underwater drones.

Sediment traps. Seston (settling particles) can be collected for subsequent analysis using traps, which vary in complexity from simple tubes open on one end to large funnels that concentrate downward particle fluxes into containers that are automatically replaced at set intervals to resolve temporal variations in sedimentation. The reliability of these techniques has been questioned, however, as they tend to suffer from artifacts that are due to sediment resuspension from the bottom, or degradation of organic material while in the trap.

Turbidity and transmissometry. Turbidity sensors use a light source and measure the light scattering by particles in the vicinity of the sensor, to assess the density of particles. Transmissometers offer a complementary view by measuring the effectiveness of the transmission of light along the length of the instrument. Both techniques are good at identifying regions in the water column with higher particle densities, including near-bottom nepheloid layers that indicate sediment resuspension.

Moored instruments. To obtain time series, instruments may be deployed for extended periods of time on moored platforms. Oceanographic moorings are often vertical cables stretched between a heavy anchor at the bottom and a buoy at or near the water surface. Instruments on moorings are commonly equipped with internal data loggers, and the recorded data series are read from them after the mooring is recovered at the end of the deployment period.

Fig. A.4. Measurements in the water column provide important information needed to interpret sediment processes. (a) Conductivity-Temperature-Depth (CTD) probe with additional instruments. (b) Rosette water sampler with a CTD mounted at the bottom. (c) Oceanographic sediment trap with a rotating array of sampling bottles. (d) Deployment diagram for a moored instrumentation array. (*Image credits*: USGS public domain (a, c, d) and this author (b).)

A.5   Incubation Experiments

While *in situ* methods remain difficult, *ex situ* sediment incubations are a common method for estimating fluxes across the sediment-water interface and reaction rates within the sediment.

Slurry incubations. These incubations use homogenized sediment, typically taken from a predefined layer within the sediment column. The Sediment is incubated in containers (or specially designed plastic bags), and changes in chemical concentrations are assessed by periodic sampling.

Whole-core incubations. These incubations use vertically-preserved sections of sediment (e.g., multicorer cores; Fig. A.5). The cores are typically fitted with caps that may have built-in controls for stirring, maintaining oxygen concentrations, and other functions, as well as sampling ports that allow periodic withdrawal of water.

Mesocosms. Larger volumes of sediment may be used in incubations that mimic conditions in the benthic ecosystem, for example, to study the behavior of benthic animals.

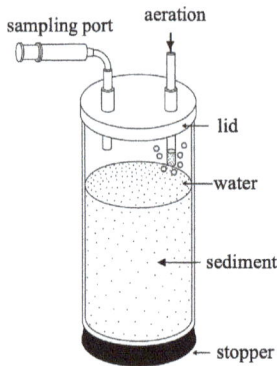

Fig. A.5.   Whole-core sediment incubation experiment.

Appendix B

Mineralogy

Table B.1. Selected diagenetically important minerals.

	Mineral	Composition	Comments
1	Calcite	$CaCO_3$	—
2	Aragonite	$CaCO_3$	—
3	Ferrihydrite	$Fe(OH)_3$	Nanophase in fresh precipitate
4	Goethite	α-FeOOH	—
5	Lepidocrocite	γ-FeOOH	—
6	Hematite	Fe_2O_3	—
7	Magnetite	Fe_3O_4	Mixed valence Fe(II, III)
8	Siderite	$FeCO_3$	Fe(II)
9	Mackinawite	FeS	—
10	Greigite	Fe_3S_4	Mixed valence Fe(II, III)
11	Pyrite	FeS_2	—
12	Burnessite	$MnO_2 \cdot nH_2O$	Non-stoichiometric, Mn(III, IV)
13	Pyrolusite	β-MnO_2	Mn(IV)
14	Manganite	γ-MnOOH	Mn(III)
15	Rhodochrosite	$MnCO_3$	Mn(II)
16	Sulfur (elemental)	S^0, S_8, S_x	Polysulfides
17	Gypsum	$CaSO_4 \cdot 2H_2O$	—
18	Barite	$BaSO_4$	—
19	Apatite	$Ca_5(PO_4)_3(F, Cl, OH)$	—
20	Vivianite	$Fe_3(PO_4)_2 \cdot 8\,H_2O$	Fe(II)

Appendix C

Chemical Equilibria in Aqueous Solutions

C.1 Equilibrium Ion Speciations

Table C.1 lists the equilibrium constants for the speciation of some common ions.

Table C.1. Equilibrium reactions. Molar (M) units should be used for concentrations if used in dilute solutions in place of activities. Values are listed at 25°C.

Reaction	K_{eq}	Value
$H_2O \longleftrightarrow H^+ + OH^-$	$K_w = [H^+][OH^-]$	1.01×10^{-14}
$CO_2(aq) + H_2O \longleftrightarrow HCO_3^- + H^+$	$K_1 = \dfrac{[HCO_3^-][H^+]}{[CO_2]}$	4.47×10^{-7}
$HCO_3^- \longleftrightarrow CO_3^{2-} + H^+$	$K_2 = \dfrac{[CO_3^{2-}][H^+]}{[HCO_3^-]}$	4.68×10^{-11}
$NH_4^+ \longleftrightarrow NH_3 + H^+$	$K_{NH_3} = \dfrac{[NH_3][H^+]}{[NH_4^+]}$	5.68×10^{-10}
$H_2S \longleftrightarrow HS^- + H^+$	$K_{H_2S} = \dfrac{[HS^-][H^+]}{[H_2S]}$	9.30×10^{-8}
$H_3PO_4 \longleftrightarrow H_2PO_4^- + H^+$	$K_{1.PO_4} = \dfrac{[H_2PO_4^-][H^+]}{[H_3PO_4]}$	6.2×10^{-8}
$H_2PO_4^- \longleftrightarrow HPO_4^{2-} + H^+$	$K_{2.PO_4} = \dfrac{[HPO_4^{2-}][H^+]}{[H_2PO_4^-]}$	6.2×10^{-8}
$HPO_4^{2-} \longleftrightarrow PO_4^{3-} + H^+$	$K_{3.PO_4} = \dfrac{[PO_4^{3-}][H^+]}{[HPO_4^{3-}]}$	3.6×10^{-13}

C.1.1 Carbonate speciation

The carbonate equilibrium in Fig. C.1 was calculated as

$$\frac{[\text{CO}_2]}{C_{\text{tot}}} = \left[1 + \frac{K_1}{[\text{H}^+]} + \frac{K_1 K_2}{[\text{H}^+]^2}\right]^{-1} \tag{C.1}$$

$$\frac{[\text{HCO}_3^-]}{C_{\text{tot}}} = \left[1 + \frac{[\text{H}^+]}{K_1} + \frac{K_2}{[\text{H}^+]}\right]^{-1} \tag{C.2}$$

$$\frac{[\text{CO}_3^{2-}]}{C_{\text{tot}}} = \left[1 + \frac{[\text{H}^+]}{K_2} + \frac{[\text{H}^+]^2}{K_1 K_2}\right]^{-1} \tag{C.3}$$

C.1.2 Ammonia speciation

The ammonia/ammonium equilibrium curves in Fig. C.2 were calculated as

$$\frac{[\text{NH}_4^+]}{[\text{NH}_3]_{\text{tot}}} = \frac{[\text{NH}_4^+]}{[\text{NH}_3] + [\text{NH}_4^+]} = \left(\frac{K_{\text{NH}_3}}{[\text{H}^+]} + 1\right)^{-1} \tag{C.4}$$

$$\frac{[\text{NH}_3]}{[\text{NH}_3]_{\text{tot}}} = \frac{[\text{NH}_3]}{[\text{NH}_3] + [\text{NH}_4^+]} = 1 - \frac{[\text{NH}_4^+]}{[\text{NH}_3] + [\text{NH}_4^+]} \tag{C.5}$$

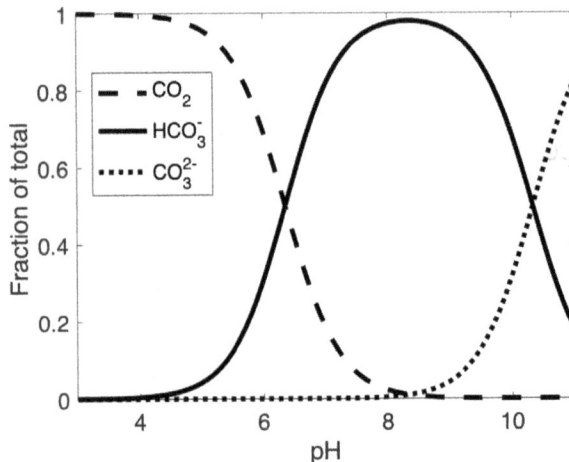

Fig. C.1. The carbonate system equilibria: Partitioning among dissolved carbon dioxide, bicarbonate ion, and carbonate ion as functions of pH.

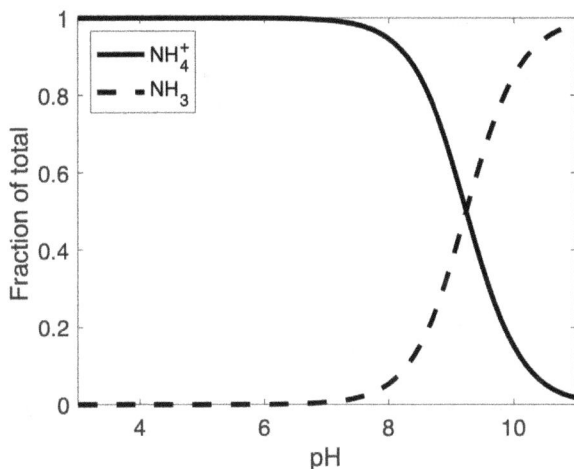

Fig. C.2. Dissociation of ammonia: Equilibrium concentrations as functions of pH.

C.1.3 Hydrogen sulfide speciation

The hydrogen sulfide equilibrium curves in Fig. C.3 were calculated as

$$\frac{[H_2S]}{[H_2S]_{tot}} = \frac{[H_2S]}{[H_2S] + [HS^-]} = \left(\frac{K_{H_2S}}{[H^+]} + 1\right)^{-1} \tag{C.6}$$

$$\frac{[HS^-]}{[H_2S]_{tot}} = \frac{[HS^-]}{[H_2S] + [HS^-]} = 1 - \frac{[H_2S]}{[H_2S] + [HS^-]} \tag{C.7}$$

C.1.4 Phosphate speciation

The phosphate equilibrium curves in Fig. C.4 were calculated as

$$\frac{[PO_4^{3-}]}{[P_{tot}]} = \left[1 + \frac{[H^+]}{K_3} + \frac{[H^+]^2}{K_2K_3} + \frac{[H^+]^3}{K_1K_2K_3}\right]^{-1} \tag{C.8}$$

$$\frac{[HPO_4^{2-}]}{[P_{tot}]} = \frac{[H^+]}{K_3}\frac{[PO_4^{3-}]}{[P_{tot}]} \tag{C.9}$$

$$\frac{[H_2PO_4^-]}{[P_{tot}]} = \frac{[H^+]^2}{K_2K_3}\frac{[PO_4^{3-}]}{[P_{tot}]} \tag{C.10}$$

$$\frac{[H_3PO_4]}{[P_{tot}]} = \frac{[H^+]^3}{K_1K_2K_3}\frac{[PO_4^{3-}]}{[P_{tot}]} \tag{C.11}$$

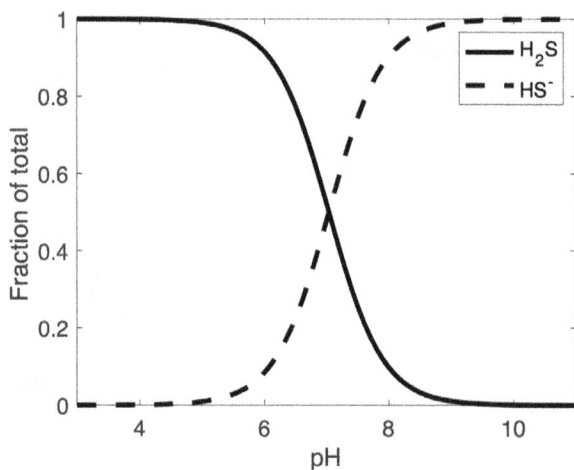

Fig. C.3. Dissociation of hydrogen sulfide: Equilibrium concentrations as functions of pH.

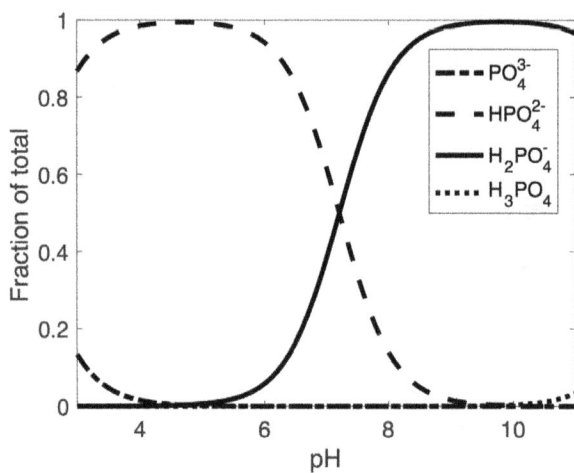

Fig. C.4. Dissociation of phosphate: Equilibrium concentrations as functions of pH.

C.2 Activity Coefficients

In solutions with an ionic strength up to 0.5 M, activity coefficients for ionic species may be calculated from the *Debye–Hückel equation*:

$$- \log \gamma = A Z^2 \sqrt{I} \qquad \text{(C.12)}$$

Table C.2. Parameters for the Debye–Hückel equation at 1 atm. From [234].

T (°C)	A
0	0.4883
5	0.4921
10	0.4960
15	0.5000
20	0.5042
25	0.5085
30	0.5131

where Z is the charge of the ion and I is the ionic strength of the solution:

$$I = \frac{1}{2} \sum_i m_i Z_i^2 \qquad (C.13)$$

where m_i are the molar concentrations for the species in the solution. The coefficients A depend on temperature and pressure. Table C.2 lists their values at 1 atm.

C.3 Solubility of Minerals

Table C.3 lists the solubility product constants for some diagenetically important minerals.

C.4 Solubility of Gases

Table C.4 lists the values of Henry's law constants K_H for some gases. At high concentrations or high pressures, C_{aq} and P in Eq. 3.1 should be replaced, respectively, by activity and fugacity.

The decrease in gas solubility with increasing temperature may be approximated [6], for small deviations from the standard temperature, using the Clausius–Clapeyron equation as

$$\ln C = \ln C_0 - \frac{\Delta H_{\text{sol}}}{R} \left(\frac{1}{T} - \frac{1}{T_0} \right) \qquad (C.14)$$

Table C.3. Solubility of selected minerals (at 25°C and 1 atm) [22, 24, 27, 125, 235].

Mineral	Reaction	$\log \Omega_{eq}$
Calcite	$CaCO_3 \longleftrightarrow Ca^{2+} + CO_3^{2-}$	-8.480
Aragonite	$CaCO_3 \longleftrightarrow Ca^{2+} + CO_3^{2-}$	-8.336
Siderite	$FeCO_3 \longleftrightarrow Fe^{2+} + CO_3^{2-}$	-10.45
Rhodocrosite	$MnCO_3 \longleftrightarrow Mn^{2+} + CO_3^{2-}$	-11.13
Gypsum	$CaSO_4 \cdot 2\,H_2O \longleftrightarrow Ca^{2+} + SO_4^{2-} + 2\,H_2O$	-4.58
Barite	$BaSO_4 \longleftrightarrow Ba^{2+} + SO_4^{2-}$	-9.97
Mackinawite	$FeS + 2\,H^+ \longleftrightarrow Fe^{2+} + H_2S$	3.5
Vivianite	$Fe_3(PO_4)_2 \cdot 8\,H_2O \longleftrightarrow 3\,Fe^{2+} + 2\,PO_4^{3-} + 8\,H_2O$	-35.8
Ferrihydrite	$Fe(OH)_3 + 3\,H^+ \longleftrightarrow Fe^{3+} + 3\,H_2O$	1.62
Lepidocrocite	$FeOOH + 3\,H^+ \longleftrightarrow Fe^{3+} + 2\,H_2O$	0.46

Table C.4. Henry's law constants. From [22, 236].

Gas	K_H (M/atm)	ΔH_{sol} (kJ/mol)
CH_4	1.3×10^{-3}	-13.4
CO_2	3.4×10^{-2}	-19.3
N_2O	2.6×10^{-2}	—
N_2	6.0×10^{-4}	—
NH_3	6.0×10^{1}	-2.0
H_2S	1.0×10^{-1}	—

where the enthalpy of solution at standard conditions ΔH_{sol} is given in Table C.4. C_0 is the saturation concentration at standard conditions.

C.5 Gibbs Free Energies of Formation

Changes in the standard Gibbs free energies of reactions ΔG^0 may be obtained from the corresponding energies of formation for the reactants and products (Table C.5). For example, formation of water

$$H^+ + OH^- \longrightarrow H_2O$$

Table C.5. Standard Gibbs free energies of formation for common molecules [22, 237, 238]. "s"=solid, "l"=liquid, "g"=gas, "aq"=aqueous (dissolved).

Compound	G_f^0 (kJ/mol)
H^+ (aq)	0
OH^- (aq)	−157.29
H_2O (l)	−237.18
H_2O (g)	−228.57
H_2 (aq)	17.57
O_2 (aq)	16.32
CO_2 (g)	−394.39
CO_2 (aq)	−386.23
HCO_3^-	−587.2
CH_4 (aq)	−34.39
SO_4^{2-}	−744.8
SO_3^{2-}	−486.6
$S_2O_3^{2-}$	−513.4
HS^-	11.97
H_2S	−27.9
H_2S (aq)	−27.87
S (aq)	85.8
S^0 (s)	0
NH_4^+	−80.12
NH_3 (aq)	−26.57
NO_2^-	−32.22
NO_3^-	−110.83
N_2 (g)	0
HPO_4^{2-} (aq)	−1089.3
$H_2PO_4^-$ (aq)	−1130.4
Fe^{2+} (aq)	−78.87
FeS (s mackinawite)	−93.30
FeS (s)	−89.12
FeS_2 (s pyrite)	−160.2
$FeOH_3$ (s ferrihydrite)	−699
FeOOH (s goethite)	−462
$Fe_3(PO_4)_2 \cdot 8\,H_2O$ (s vivianite)	−4377.2
Mn^{2+} (aq)	−228.0
MnO_2 (s pyrolusite)	−465.1
Glucose	−915.9
Lactate	−517.8
Acetate	−369.32
Pyruvate	−474.6

corresponds to the free energy difference of $(-237.18) - [0 + (-158.29)] = -79.89$ kJ/mol. The negative sign indicates that the energy is released in the forward reaction. Note that this thermodynamic favorability (meaning the reaction is not in equilibrium) was calculated for the standard concentrations of 1 M. At the OH^- and H^+ concentrations of 10^{-7} M each (corresponding to pH 7), Eq. (3.4) indicates that the energy difference ΔG becomes zero and the reaction reaches equilibrium.

C.6   Redox Half Reactions

Table C.6 lists selected half-reactions in their reduction form. Oxidation half-reactions can be obtained by reversing the reaction equation and changing the sign of the E_H^0. The relationship between ΔG^0 and E_H^0 is given by Eq. (3.15): The E_H^0 effectively describes the Gibbs free energy change per electron (with the Faraday constant F serving as the unit conversion factor).

Table C.6.   Selected redox half reactions [6, 15].

Oxidized form		Reduced form	E_H^0 (V) (standard)	E_H^0 (V) (pH 7)
$O_2 + 4\,H^+ + 4\,e^-$	$\longrightarrow$	$2\,H_2O$	1.27	0.89
$NO_3^- + 6\,H^+ + 5\,e^-$	$\longrightarrow$	$\frac{1}{2}\,N_2(g) + 3\,H_2O$	1.24	0.78
$NO_3^- + 10\,H^+ + 8\,e^-$	$\longrightarrow$	$NH_4^+ + 3\,H_2O$	0.88	0.40
$NO_3^- + 2\,H^+ + 2\,e^-$	$\longrightarrow$	$NO_2^- + H_2O$	0.85	0.47
$NO_2^- + 4\,H^+ + 3\,e^-$	$\longrightarrow$	$\frac{1}{2}\,N_2(g) + 2\,H_2O$	1.53	1.02
$NO_2^- + 8\,H^+ + 6\,e^-$	$\longrightarrow$	$NH_4^+ + 2\,H_2O$	0.90	0.65
$MnO_2 + 4\,H^+ + 2\,e^-$	$\longrightarrow$	$Mn^{2+} + 2\,H_2O$	1.23	0.46
$FeOOH + 3\,H^+ + e^-$	$\longrightarrow$	$Fe^{2+} + 2\,H_2O$	0.95	−0.21
$Fe(OH)_3 + 3\,H^+ + e^-$	$\longrightarrow$	$Fe^{2+} + 3\,H_2O$	0.95	−0.20
$SO_4^{2-} + 10\,H^+ + 8\,e^-$	$\longrightarrow$	$H_2S + 4\,H_2O$	0.30	−0.18
$S^0(s) + 2\,H^+ + 2\,e^-$	$\longrightarrow$	H_2S	0.14	−0.24
$S_2O_3^{2-} + 10\,H^+ + 8\,e^-$	$\longrightarrow$	$2\,H_2S$	0.33	−0.15
$HCO_3^- + 9\,H^+ + 8\,e^-$	$\longrightarrow$	$CH_4 + 3\,H_2O$	0.21	−0.22
$HCO_3^- + 5\,H^+ + 4\,e^-$	$\longrightarrow$	$\frac{1}{6}\,C_6H_{12}O_6 + 2\,H_2O$	0.20	−0.22
$2\,HCO_3^- + 9\,H^+ + 8\,e^-$	$\longrightarrow$	$C_2H_3O_2^- + 4\,H_2O$	0.19	−0.25
$2\,H^+ + 2\,e^-$	$\longrightarrow$	$H_2(g)$	0.00	−0.38

The E_H^0 values at standard conditions can be obtained from the Gibbs free energies of formation in Table C.5, using Eq. (3.15). The values for the environmentally relevant conditions of pH 7 were calculated in Table C.6 as

$$E_H^0(pH7) = E_H^0 - \frac{RT}{nF} \ln [H^+]^\nu = E_H^0 - \frac{RT}{nF} 7|\nu| \ln 10 \qquad (C.15)$$

where ν is the (negative) stoichiometric coefficient in the corresponding half-reaction equation and n is the number of electrons in that reaction. The calculation used the temperature of $T = 4°C$, rather than the standard temperature of 25°C, to better represent conditions in typical deep water sediments. The listed E_H^0 values still assume the concentrations of other solutes at 1 M, and may require corrections to reflect the actual concentrations in the environment (see Section 3.4.3).

Appendix D

Common Reactions

D.1 Reactions

Table D.1. Selected biogeochemical reactions*. Gibbs free energies ΔG^0 (kJ/mol) were calculated at standard conditions, for the reactions as written. Environmentally relevant values ΔG correspond to pH 7, $T = 4°C$, concentrations of HCO_3^- of 1 mM, and concentrations of all other dissolved species of 10 μM.

No.	Reaction	ΔG^0	ΔG
	Primary redox reactions		
1	$CH_2O + O_2 \longrightarrow HCO_3^- + H^+$	-451	-451
2	$CH_2O + \frac{4}{5} NO_3^- \longrightarrow HCO_3^- + \frac{2}{5} N_2 + \frac{2}{5} H_2O + \frac{1}{5} H^+$	-441	-416
3	$CH_2O + 2\,MnO_2 + 3\,H^+ \longrightarrow HCO_3^- + 2\,Mn^{2+} + 2\,H_2O$	-435	-366
4	$CH_2O + 4\,FeOOH + 7\,H^+ \longrightarrow HCO_3^- + 4\,Fe^{2+} + 6\,H_2O$	-376	-212
5	$CH_2O + \frac{1}{2} SO_4^{2-} \longrightarrow HCO_3^- + \frac{1}{2} H_2S$	-76	-65
	Methanogenesis		
6†	$CH_2O + \frac{1}{2} H_2O \longrightarrow \frac{1}{2} HCO_3^- + \frac{1}{2} CH_4 + \frac{1}{2} H^+$		
7	$CH_3COO^- + H_2O \longrightarrow HCO_3^- + CH_4$	-15	-31
8	$CO_2 + 4\,H_2 \longrightarrow CH_4 + 2\,H_2O$	-194	-61
	Secondary redox reactions		
9	$4\,Fe^{2+} + O_2 + 10\,H_2O \longrightarrow 4\,Fe(OH)_3 + 8\,H^+$	-73	-238
10‡	$Mn^{2+} + \frac{1}{2} O_2 + H_2O \longrightarrow MnO_2 + 2\,H^+$	-8	-42
11	$H_2S + 2\,O_2 + 2\,H_2O \longrightarrow SO_4^{2-} + 2\,H^+$	-275	-296
12	$NH_4^+ + 2\,O_2 \longrightarrow NO_3^- + \frac{2}{5} H_2O + 2\,H^+$	-158	-179
13	$CH_4 + 2\,O_2 \longrightarrow CO_2 + 2\,H_2O$	-859	-832

(*Continued*)

Table D.1. (*Continued*)

No.	Reaction	ΔG^0	ΔG
14	$2\,Fe^{2+} + MnO_2 + 4\,H_2O \longrightarrow 2\,Fe(OH)_3 + Mn^{2+} + 2\,H^+$	-29	-76
15	$10\,Fe^{2+} + 2\,NO_3^- + 24\,H_2O \longrightarrow 10\,Fe(OH)_3 + N_2 + 18\,H^+$	-158	-509
16	$4\,Fe^{2+} + 2\,NO_2^- + 5\,H_2O \longrightarrow 4\,FeOOH + N_2O + 6\,H^+$	-127	-217
17	$2\,Fe(OH)_3 + H_2S + 4\,H^+ \longrightarrow 2\,Fe^{2+} + S^0 + 6\,H_2O$	-181	-59
18	$SO_4^{2-} + CH_4 \longrightarrow HS^- + HCO_3^- + H_2O$	-33	-23
19	$8\,FeOOH + CH_4 + 15\,H^+ \longrightarrow 8\,Fe^{2+} + HCO_3^- + 13\,H_2O$	-674	-318
	Hydrogenotrophic respiratory reactions		
20	$O_2 + 2\,H_2 \longrightarrow 2\,H_2O$	-526	-447
21	$MnO_2 + H_2 + 2\,H^+ \longrightarrow Mn^{2+} + 2\,H_2O$	-255	-181
22	$2\,Fe(OH)_3 + H_2 + 4\,H^+ \longrightarrow 2\,Fe^{2+} + 6\,H_2O$	-226	-105
23	$S^0 + H_2 \longrightarrow H_2S$	-46	-46
24	$SO_4^{2-} + 4\,H_2 + 2\,H^+ \longrightarrow H_2S + 4\,H_2O$	-303	-123
25	$2\,HCO_3^- + 4\,H_2 + H^+ \longrightarrow CH_3COO^- + 4\,H_2O$	-215	-66
	Disproportionation		
26	$4\,S^0 + 4\,H_2O \longrightarrow SO_4^{2-} + 3\,HS^- + 5\,H^+$	$+240$	-51
	Mineral precipitation reactions		
27	$Fe^{2+} + HS^- \longrightarrow FeS(s) + H^+$		
28	$FeS + H_2S \longrightarrow FeS_2 + H_2$		
29	$FeS + S^0 \longrightarrow FeS_2$		
30	$3\,Fe^{2+} + 2\,H_2PO_4^- \longrightarrow Fe_3(PO_4)_2 + 4\,H^+$		

Notes: [*]For consistency, dissolved CO_2 is represented using the carbonate ion HCO_3^-, which is the dominant ion at circumneutral pH. For alternative formulations of the carbonate balance, one can use, for example, relationships such as $CO_2 + H_2O \longrightarrow HCO_3^- + H^+$ and rearrange the equations.
[†]Nominal reaction using generic stoichiometry of organic matter. See reaction 7 for the specific case of acetoclastic methanogenesis.
[‡]Typically surface-catalyzed.

D.2 Reaction kinetics

Table D.2. Selected rate formulations and reaction rate constants.
Note that the values of rate constants may vary among different environments by
several orders of magnitude and may depend on pH, salinity, and other variables.
Values in this table are provided for guidance only [206]. Multiple alternative
formulations of the rate laws are possible, especially for the microbially-mediated
reactions that may exhibit Monod kinetics. Numbers in parentheses next to the
reaction names correspond to the row numbers in Table D.1. Time scales are
stated for typical concentrations within reaction zones: solutes on the order of
10 μM and solids on the order of 10 μmol/cm^{-3}.

Rate law	k		Time scale
OC mineralization (1–6) (Fig. 5.6)			
$R = k[\text{OC}]$	10^{-5}–10^{1}	y^{-1}	weeks to millennia
Aerobic oxidation of iron (9) [208] [239]†			
$R = k[\text{Fe}^{2+}][\text{O}_2]$	10^{2}	μM^{-1}y^{-1}	hours to days
$R = k[\text{Fe}^{2+}][\text{O}_2][\text{H}^+]^{-2}$	1	μM y^{-1}	
Kinetics is strongly pH-dependent and facilitated by Fe oxyhydroxides.			
Aerobic oxidation of sulfide (11) [240, 241]† [159, 242]‡			
$R = k[\text{H}_2\text{S}][\text{O}_2]$	10^{0}–10^{3}	μM^{-1}y^{-1}	hours to weeks
Kinetics is linked to the concentration of HS$^-$ and is thus pH-dependent [240].			
Aerobic oxidation of ammonium (12) [208, 243]‡			
$R = k[\text{NH}_4^+][\text{O}_2]$	10^{1}	μM^{-1}y^{-1}	days to weeks
Aerobic oxidation of methane (13) [208]‡ [244]			
$R = k[\text{CH}_4][\text{O}_2]$	10^{-1} 10^{4}	μM^{-1}y^{-1}	days
Fe(III) reduction by H$_2$S (17) [245]			
$R = k[\text{Fe(III)}][\text{H}_2\text{S}]$	10^{-2}	μM^{-1}y^{-1}	months
S-dependent anaerobic oxidation of methane (18) [246, 247]‡			
$R = k[\text{SO}_4^{2-}][\text{CH}_4]$	10^{-5}	μM^{-1}y^{-1}	centuries
Fe-dependent anaerobic oxidation of methane (19) [248]‡			
$R = k[\text{Fe(III)}][\text{CH}_4]$	10^{-4}–10^{-3}	μM^{-1}y^{-1}	decades
Elemental sulfur disproportionation (26) [159, 201, 217]‡			
$R = k[\text{S}^0]$	10^{-1}–10^{1}	y^{-1}	weeks to years

Notes: †determined in an abiotic system.
‡model-constrained.

Appendix E

Stable and Radioactive Isotopes

Each chemical element in the periodic table is defined by the number of protons in its atoms. The atom of carbon, for example, has 6 protons, so it is found in the sixth position in the periodic table. The number of neutrons in each atom is approximately equal to the number of protons, but their exact number may differ slightly. Carbon atoms, for example, may have six, seven, or eight neutrons. Atoms of the same element that have different numbers of neutrons are referred to as *isotopes*. Different isotopes of the same element are denoted by the sum of their protons and neutrons. The isotopes of carbon, therefore, are written as ^{12}C, ^{13}C and ^{14}C. Because the masses of neutrons and protons are nearly equal, the superscript approximately corresponds to the *atomic mass*. One mole of ^{12}C weighs 12 grams.

E.1 Radioactive Isotopes

For some numbers of neutrons, atomic nuclei are unstable. Over time, such atoms decay, generating new (lighter) chemical elements. Such unstable isotopes are called *radioactive*. The probability of decay for any given atom is governed by the laws of quantum mechanics and remains constant over time. The number of atoms in a sample that decay during any given time interval is therefore proportional to the total number of atoms:

$$\frac{dN}{dt} = -\gamma N \qquad (E.1)$$

where γ characterizes the probability of decay. Integrating this equation gives the well-known equation of an exponential decay:

$$N(t) = N_0 e^{-\gamma t} \tag{E.2}$$

Starting with the initial number N_0, the number of remaining atoms decreases with time exponentially.

Non-radioactive isotopes are called *stable*.[1] For carbon, the isotopes ^{12}C and ^{13}C are stable, whereas ^{14}C is radioactive. After every 5,730-year period, only half of the initial number of ^{14}C atoms remains. This characteristic length of time ($\tau_{1/2} = -\ln 0.5/\gamma = 0.6931/\gamma$) is referred to as the *half-life* of the radioactive isotope (Table E.1).

E.2   Stable Isotopes

Stable isotopes with the smallest number of neutrons are typically the most abundant (Table E.2). For example, 98.93% of naturally occurring carbon is ^{12}C. The abundances of other stable isotopes are quantified by *isotope ratios*, which are defined as ratios of the abundance of heavier isotope to the abundance of the lighter (most common) isotope, e.g.:

$$R = \frac{^{13}C}{^{12}C}$$

For historical reasons, these simple ratios are rarely used. The instruments (mass spectrometers) that determine isotopic abundances are much better at measuring differences between isotope ratios than the ratios themselves. The ratios are therefore measured relative to a standard material, for which the ratio of the isotopes is known. Because these differences are small, the results are traditionally expressed in parts per thousand (per mil). This is the so-called *delta notation*:

$$\delta = \frac{R_{\text{sample}} - R_{\text{std}}}{R_{\text{std}}} \times 1000 \, \text{‰}$$

---

[1]From a quantum-mechanical point of view, no isotope is truly stable. But for some, the decay times are greater than the age of the Universe, so they are effectively stable for all practical purposes.

Table E.1. Radioactive isotopes.

Isotope	Half life	Type of decay
^{3}H	12.32 y	beta
^{7}Be	53.3 d	electron capture
^{14}C	5730 y	beta
^{35}S	87.37 d	beta
^{55}Fe	2.73 y	electron capture
^{90}Sr	28.8 y	beta
^{137}Cs	30.05 y	beta, gamma
^{210}Pb	22.3 y	beta

Table E.2. Stable isotopes and their standards. Numbers in parentheses are natural abundances.

Reference	Stable	Abundance (%)			Standards
^{1}H (99.98%)	^{2}H (D)	0.02			VSMOW, SLAP
^{16}O (99.76%)	^{17}O ^{18}O	0.04	0.20		VSMOW, SLAP
^{12}C (98.9%)	^{13}C	1.1			VPDB
^{14}N (99.6%)	^{15}N	0.4			N$_2$ in air
^{32}S (94.99%)	^{33}S ^{34}S ^{36}S	0.75	4.25	0.01	CDT
^{54}Fe (5.85%)	^{56}Fe ^{57}Fe ^{58}Fe	91.75	2.12	0.28	IRMM

While it is somewhat complicated, the notation has a number of advantages. For example, when equal amounts of materials with δ^{13}C values of 2‰ and 4‰ are mixed, the resultant mixture will have a δ^{13}C of 3‰.

E.3 Isotopic Fractionations

Chemical reactions, including those mediated by microbes, typically proceed more slowly for heavier isotopes. Heavier atoms move slower for the same amount of thermal energy, and have lower vibrational frequencies in chemical bonds. Such differences generate *isotopic fractionation*: the products of the reaction receive a higher proportion of the lighter isotopes, causing them to have lower δ values than the original pool of reactants. The reactants pool becomes correspondingly enriched in the remaining heavier isotopes, acquiring higher δ values.

Isotopic fractionation for a given reaction is described by the *fractionation factor*, α, which defines the proportion of the isotope in the reaction product relative to that in the original pool. For instance, if the ratio R in the newly produced molecules of the product is 2.7% higher than in the reactant, then $\alpha = 1.027$. The *isotopic fractionation* is $\epsilon = \alpha - 1$, commonly expressed in per mil. Thus, the isotopic fractionation factor $\alpha = 1.027$ corresponds to the isotopic fractionation of $\epsilon = 27\permil$, and the δ values of the product will be 27‰ lower than those of the reactant.

When a fractionating reaction happens in a pool of molecules that is not replenished from the outside (i.e., in a closed system), the preferential removal of lighter isotopes causes the reactant pool to gradually become isotopically heavier. With constant fractionation, this makes the products of the reaction also become progressively heavier. Over time, as the reactants are consumed, the isotopic composition of the product in a completely closed system should approach that of the initial reactant (as all atoms, both light and heavy, are eventually transferred into the product). This process is called *Rayleigh distillation*. For example, in sediments at steady state, the δ-value of the reactants may be increasing with depth, and the δ-value of the products would be also increasing, trending towards the δ-value of the reactant at the sediment surface. In systems with partial exchange of atoms with the outside environment, distillation may be partial.

Fractionations are temperature-dependent and also depend on whether the process is controlled by thermodynamics or kinetics [249]. *Equilibrium fractionations* occur when isotopes are exchanged between two phases in a reaction that is close to equilibrium. The ratios do not change in time, but they may be different between the two phases. The fractionation factor α in this case can be defined as the ratio of isotopes in the two phases. For example, when calcium carbonate is in equilibrium exchanging oxygen atoms with ambient water, the carbonate phase is enriched in ^{18}O relative to H_2O, with

$$\alpha_{(CaCO_3 - H_2O)} = \frac{R_{CaCO_3}}{R_{H_2O}} = \frac{(^{18}O/^{16}O)_{CaCO_3}}{(^{18}O/^{16}O)_{H_2O}} = 1.031$$

(at 25°C), corresponding to an isotopic fractionation of 31‰.

Kinetic fractionations refer to reactions far from thermodynamic equilibrium. Fractionation arises because different isotopes undergo transformations at different rates. The fractionation factor is then equivalent to the ratio of the respective reaction rate constants:

$$\alpha = k_{\text{light}}/k_{\text{heavy}}$$

Ratios greater than 1 correspond to the lighter isotope being transformed preferentially relative to the heavier one. Kinetic fractionations are generally different from thermodynamic fractionations for the same reaction. Accordingly, fractionations for the same process may vary depending on how far the reaction is from equilibrium, which depends on the concentrations of the reactants and products and, for microbial processes, on cell-specific reaction rates [250].

Bibliography

[1]   R. A. Berner. *Early diagenesis: A Theoretical Approach.* Princeton University Press, 1980.

[2]   B. P. Boudreau. *Diagenetic Models and Their Implementation.* Springer, Berlin, 1997.

[3]   D. J. Burdige. *Geochemistry of Marine Sediments.* Princeton University Press, Princeton, 2006.

[4]   K. O. Konhauser. *Introduction to Geomicrobiology.* John Wiley & Sons, 2009.

[5]   C. M. Bethke. *Geochemical and Biogeochemical Reaction Modeling.* Cambridge University Press, 2nd edition, 2008.

[6]   D. E. Canfield, E. Kristensen, and B. Thamdrup. *Aquatic Geomicrobiology.* Elsevier, 2005.

[7]   D. E. Canfield. The early history of atmospheric oxygen: Homage to Robert M. Garrels. *Annu. Rev. Earth Planet. Sci,* 33:1–36, 2005.

[8]   E. D. Swanner, N. Lambrecht, C. Wittkop, C. Harding, S. Katsev, J. Torgeson, and S. W. Poulton. The biogeochemistry of ferruginous lakes and past ferruginous oceans. *Earth-Sci. Rev.,* 211:103430, 2020.

[9]   C. Darwin. *The Formation of Vegetable Mould, Through the Action of Worms: With Observations on Their Habits.* J. Murray, 1892.

[10] F. J. R. Meysman, J. J. Middelburg, and C. H. R. Heip. Bioturbation: A fresh look at Darwin's last idea. *Trends Ecol. Evol.,* 21(12):688–695, 2006.

[11] J. Li, S. A. Crowe, D. Miklesh, M. Kistner, D. E. Canfield, and S. Katsev. Carbon mineralization and oxygen dynamics in sediments with deep oxygen penetration, Lake Superior. *Limnol. Oceanogr.,* 57(6):1634–1650, 2012.

[12] F. J. R. Meysman, B. P. Boudreau, and J. J. Middelburg. Modeling reactive transport in sediments subject to bioturbation and compaction. *Geochim. Cosmochim. Acta*, 69(14):3601–3617, 2005.

[13] W. Hayduk and H. Laudie. Prediction of diffusion coefficients for nonelectrolytes in dilute aqueous solutions. *AIChE J.*, 20(3):611–615, 1974.

[14] E. Kristensen, G. Penha-Lopes, M. Delefosse, T. Valdemarsen, C. O. Quintana, and G. T. Banta. What is bioturbation? The need for a precise definition for fauna in aquatic sciences. *Mar. Ecol. Prog. Ser.*, 446:285–302, 2012.

[15] R. C. Aller. Sedimentary diagenesis, depositional environments, and benthic fluxes. *Treatise on Geochemistry: Second Edition*, 8:293–334, 2014.

[16] F. J. R. Meysman, B. P. Boudreau, and J. J. Middelburg. Relations between local, nonlocal, discrete and continuous models of bioturbation. *J. Mar. Res.*, 61(3):391–410, 2003.

[17] O. Maire, P. Lecroart, F. Meysman, R. Rosenberg, J. Duchêne, and A. Grémare. Quantification of sediment reworking rates in bioturbation research: A review. *Aquat. Biol.*, 2(3):219–238, 2008.

[18] I. P. Dwyer, D. A. S. Perger, M. Graffam, R. C. Aller, L. M. Wehrmann, and N. Volkenborn. Seasonality of bioirrigation by the maldanid polychaete clymenella torquata and related oxygen dynamics in permeable sediments. *J. Exp. Mar. Biol. Ecol.*, 572:151987, 2024.

[19] F. Wenzhöfer and R. N. Glud. Small-scale spatial and temporal variability in coastal benthic o2 dynamics: Effects of fauna activity. *Limnol. Oceanogr.*, 49(5):1471–1481, 2004.

[20] C. Meile, P. Berg, P. V. Cappellen, and K. Tuncay. Solute-specific pore water irrigation: Implications for chemical cycling in early diagenesis. *J. Mar. Res.*, 63(3):601–621, 2005.

[21] P. J. leB. Williams. Meters, kilograms, seconds, but no bomb units. A zero tolerance approach to units. (With apologies to Lynne Truss). *Limnol. Oceanogr. Bull.*, 13(2):29–32, 2004.

[22] W. Stumm and J. J. Morgan. *Aquatic Chemistry: Chemical Equilibria and Rates in Natural Waters*. Wiley, 2012.

[23] J. Schieber. *Iron Sulfide Formation*, pp. 486–502. Springer Netherlands, Dordrecht, 2011.

[24] D. Rickard. The solubility of FeS. *Geochim. Cosmochim. Acta*, 70(23):5779–5789, 2006.

[25] F. H. Stillinger. Proton transfer reactions and kinetics in water. *Theor. Chem.*, 3:177–234, 1978.

[26] J. Qusheng and C. M. Bethke. Predicting the rate of microbial respiration in geochemical environments. *Geochim. Cosmochim. Acta*, 69(5):1133–1143, 2005.

[27] S. Bonneville, P. Van Cappellen, and T. Behrends. Microbial reduction of iron(III) oxyhydroxides: Effects of mineral solubility and availability. *Chem. Geol.*, 212(3-4):255–268, 2004.

[28] B. B. Jørgensen and S. Kasten. Sulfur cycling and methane oxidation. In *Marine Geochemistry*, pp. 271–309. Springer, 2006.

[29] N. Iversen and B. B. Jorgensen. Anaerobic methane oxidation rates at the sulfate-methane transition in marine sediments from Kattegat and Skagerrak (Denmark) 1. *Limnol. Oceanogr.*, 30(5):944–955, 1985.

[30] A. W. Dale, P. Regnier, and P. Van Cappellen. Bioenergetic controls on anaerobic oxidation of methane (AOM) in coastal marine sediments: A theoretical analysis. *Am. J. Sci.*, 306(4):246–294, 2006.

[31] J. Li and S. Katsev. Nitrogen cycling in deeply oxygenated sediments: Results in lake superior and implications for marine sediments. *Limno. Oceanogr.*, 59(2), 2014.

[32] M. Fakhraee, J. Li, and S. Katsev. Significant role of organic sulfur in supporting sedimentary sulfate reduction in low-sulfate environments. *Geochim. Cosmochimi. Acta*, 213:502–516, 2017.

[33] H. N. Schulz, T. Brinkhoff, T. G. Ferdelman, M. H. Mariné, A. Teske, and B. B. Jørgensen. Dense populations of a giant sulfur bacterium in Namibian shelf sediments. *Science*, 284(5413):493–495, 1999.

[34] R. Dermott and M. Legner. Dense Mat-forming bacterium thioploca ingrica (Beggiatoaceae) in eastern Lake Ontario: Implications to the benthic food web. *J. Great Lakes Res.*, 28(4):688–697, 2002.

[35] B. B. Jørgensen. Shrinking majority of the deep biosphere. *Proc. Natl. Acad. Sci. U.S.A.*, 109(40):15976–15977, 2012.

[36] U. Von Stockar, T. Maskow, J. Liu, I. W. Marison, and R. Patiño. Thermodynamics of microbial growth and metabolism: An analysis of the current situation. *J. Biotechnol.*, 121:517–533, 2006.

[37] B. B. Jørgensen. Bacteria and marine biogeochemistry. *Marine Geochemistry*, pp. 173–207, 2000.

[38] P. Noll, L. Lilge, R. Hausmann, and M. Henkel. Modeling and exploiting microbial temperature response. *Processes*, 8(1):121, 2020.

[39] Q. Jin and M. F. Kirk. pH as a Primary control in environmental microbiology: 2. Kinetic perspective. *Front. Environ. Sci.*, 6:101, 2018.

[40] C. M. Bethke, R. A. Sanford, M. F. Kirk, Q. Jin, and T. M. Flynn. The thermodynamic ladder in geomicrobiology. *Am. J. Sci.*, 311(3):183–210, 2011.

[41] D. Fernández-Calviño and E. Bååth. Growth response of the bacterial community to pH in soils differing in pH. *FEMS Microbiol. Ecol.*, 73(1):149–156, 2010.

[42] L. C. Wunder, D. A. Aromokeye, X. Yin, T. Richter-Heitmann, G. Willis-Poratti, A. Schnakenberg, C. Otersen, I. Dohrmann, M. Römer, G. Bohrmann *et al.*, Iron and sulfate reduction structure microbial communities in (sub-) antarctic sediments. *ISME J.*, 15(12):3587–3604, 2021.

[43] C. Schauberger, R. N. Glud, B. Hausmann, B. Trouche, L. Maignien, J. Poulain, P. Wincker, S. Arnaud-Haond, F. Wenzhöfer, and B. Thamdrup. Microbial community structure in hadal sediments: High similarity along trench axes and strong changes along redox gradients.*ISME J.*, 15(12):3455–3467, 2021.

[44] S. Louca, A. K. Hawley, S. Katsev, M. Torres-Beltran, M. P. Bhatia, S. Kheirandish, C. C. Michiels, D. Capelle, G. Lavik, M. Doebeli, S. A. Crowe, and S. J. Hallam. Integrating biogeochemistry with multiomic sequence information in a model oxygen minimum zone. *Proc. Natl. Acad. Sci. U.S.A.*, 113(40):E5925–E5933, 2016.

[45] K. Lepot. Signatures of early microbial life from the Archean (4 to 2.5 Ga) eon. *Earth Sci. Rev.*, 209, 103296, 2020.

[46] J. Rullkötter. Organic matter: The driving force for early diagenesis. In *Marine Geochemistry*, pp. 125–168. Springer, 2006.

[47] R. E. Hecky, P. Campbell, and L. L. Hendzel. The stoichiometry of carbon, nitrogen, and phosphorus in particulate matter of lakes and oceans. *Limnol. Oceanogr.*, 38(4):709–724, 1993.

[48] P. N. Froelich, G. P. Klinkhammer, M. L. Bender, N. A. Luedtke, G. Ross Heath, D. Cullen, P. Dauphin, D. Hammond, B. Hartman, and V. Maynard. Early oxidation of organic matter in pelagic sediments of the eastern equatorial Atlantic: Suboxic diagenesis. *Geochim. Cosmochim. Acta*, 43(7):1075–1090, 1979.

[49] R. N. Glud. Oxygen dynamics of marine sediments. *Mar. Biol. Res.*, 4(4):243–289, 2008.

[50] S. Katsev, B. Sundby, and A. Mucci. Modeling vertical excursions of the redox boundary in sediments: Application to deep basins of the Arctic Ocean. *Limnol. Oceanogr.*, 51(4), 2006.

[51] J. Li. *Sediment Diagenesis in Large Lakes Superior and Malawi, Geochemical Cycles and Budgets and Comparisons to Marine Sediments*. PhD thesis, University of Minnesota, 2014.

[52] P. Jourabchi, P. Van Cappellen, and P. Regnier. Quantitative interpretation of pH distributions in aquatic sediments: A reaction-transport modeling approach. *Am. Sci.*, 305(9):919–956, 2005.

[53] P. Jourabchi, C. Meile, L. R. Pasion, and P. Van Cappellen. Quantitative interpretation of pore water O2 and pH distributions in deep-sea sediments. *Geochim. Cosmochim. Acta*, 72(5):1350–1364, 2008.

[54] F. J. R. Meysman, N. Risgaard-Petersen, S. Y. Malkin, and L. P. Nielsen. The geochemical fingerprint of microbial long-distance electron transport in the seafloor. *Geochim. Cosmochim. Acta*, 152:122–142, 2015.

[55] K. E. A Segarra, F. Schubotz, V. Samarkin, M. Y. Yoshinaga, K. Uwe Hinrichs, and S. B. Joye. High rates of anaerobic methane oxidation in freshwater wetlands reduce potential atmospheric methane emissions. *Nat. Commun.*, 6(1):1–8, 2015.

[56] E. J. Beal, C. H. House, and V. J. Orphan. Manganese-and iron-dependent marine methane oxidation. *Science*, 325(5937):184–187, 2009.

[57] K. F. Ettwig, B. Zhu, D. Speth, J. T. Keltjens, M. S. M. Jetten, and B. Kartal. Archaea catalyze iron-dependent anaerobic oxidation of methane. *Proc. Natl. Acad. Sci.*, 113(45):12792–12796, 2016.

[58] K. Á. Norði and B. Thamdrup. Nitrate-dependent anaerobic methane oxidation in a freshwater sediment. *Geochim. Cosmochi. Acta*, 132:141–150, 2014.

[59] C. U. Welte, O. Rasigraf, A. Vaksmaa, W. Versantvoort, A. Arshad, H. J. M. Op den Camp, M. S. M. Jetten, C. Lüke, and J. Reimann. Nitrate-and nitrite-dependent anaerobic oxidation of methane. *Environ. Microbiol. Rep.*, 8(6):941–955, 2016.

[60] L. Liu, J. Wilkinson, K. Koca, C. Buchmann, and A. Lorke. The role of sediment structure in gas bubble storage and release. *J. Geophys. Res. Biogeosci*, 121(7):1992–2005, 2016.

[61] B. P. Boudreau, B. S. Gardiner, and B. D. Johnson. Rate of growth of isolated bubbles in sediments with a diagenetic source of methane. *Limnol. Oceanogr.*, 46(3):616–622, 2001.

[62] I. Ostrovsky, D. F. McGinnis, L. Lapidus, and W. Eckert. Quantifying gas ebullition with echosounder: The role of methane transport by bubbles in a medium-sized lake. *Limnol. Oceanogr. Methods*, 6(2):105–118, 2008.

[63] N. Lambrecht, S. Katsev, C. Wittkop, S. J. Hall, C. S. Sheik, A. Picard, M. Fakhraee, and E. D. Swanner. Biogeochemical and

physical controls on methane fluxes from two ferruginous meromictic lakes. *Geobiology*, 18(1):54–69, 2020.

[64]   D. Bastviken, L. J. Tranvik, J. A. Downing, P. M. Crill, and A. Enrich-Prast. Freshwater methane emissions offset the continental carbon sink. *Science*, 331(6013):50, 2011.

[65]   T. Langenegger, D. Vachon, D. Donis, and D. F. McGinnis. What the bubble knows: Lake methane dynamics revealed by sediment gas bubble composition. *Limnol. Oceanogr.*, 64(4):1526–1544, 2019.

[66]   T. DelSontro, D. F McGinnis, S. Sobek, I. Ostrovsky, and B. Wehrli. Extreme methane emissions from a swiss hydropower reservoir: Contribution from bubbling sediments. *Environ. Sci. Technol.*, 44(7):2419–2425, 2010.

[67]   E. Suess. Particulate organic carbon flux in the oceans—surface productivity and oxygen utilization. *Nature*, 288(5788):260–263, 1980.

[68]   J. Li, E. T. Brown, S. A. Crowe, and S. Katsev. Sediment geochemistry and contributions to carbon and nutrient cycling in a deep meromictic tropical lake: Lake Malawi (East Africa). *J. Great. Lakes. Res.*, 44(6): 1221–1234, 2018.

[69]   D. J. Burdige. Preservation of organic matter in marine sediments: Controls, mechanisms, and an imbalance in sediment organic carbon budgets? *Chem. Rev.*, 107(2):467–485, 2007.

[70]   S. Arndt, B. B. Jørgensen, D. E. LaRowe, J. J. Middelburg, R. D. Pancost, and P. Regnier. Quantifying the degradation of organic matter in marine sediments: A review and synthesis. *Earth-Sci. Rev.*, 123:53–86, 2013.

[71]   J. J. Middelburg, T. Vlug, F. Jaco, and W. A. van der Nat. Organic matter mineralization in marine systems. *Glob. Planet. Change*, 8(1-2):47–58, 1993.

[72]   S. Katsev and S. A. Crowe. Organic carbon burial efficiencies in sediments: The power law of mineralization revisited. *Geology*, 43(7): 607–610, 2015.

[73]   B. B. Jørgensen, B. B. Jørgensen, and R. J. Parkes. Role of sulfate reduction and methane production by organic carbon degradation in eutrophic fjord sediments (Limfjorden, Denmark). *Limnol. Oceanogr.*, 55(3):1338–1352, 2010.

[74]   J. Li, E. T. Brown, S. A. Crowe, and S. Katsev. Sediment geochemistry and contributions to carbon and nutrient cycling in a deep meromictic tropical lake: Lake Malawi (East Africa). *J. Great. Lakes. Res.*, 44(6):1221–1234, 2018.

[75]   R. C. Aller. Mobile deltaic and continental shelf muds as suboxic, fluidized bed reactors. *Mar. Chem.*, 61(3-4):143–155, 1998.

[76] L. J. Tranvik, J. A. Downing, J. B. Cotner, S. A. Loiselle, R. G. Striegl, T. J. Ballatore, P. Dillon, K. Finlay, K. Fortino, L. B. Knoll, P. L. Kortelainen, T. Kutser, S. Larsen, I. Laurion, D. M. Leech, S. L. McCallister, D. M. McKnight, J. M. Melack, E. Overholt, J. A. Porter, Y. Prairie, W. H. Renwick, F. Roland, B. S. Sherman, D. W. Schindler, S. Sobek, A. Tremblay, M. J. Vanni, A. M. Verschoor, E. von Wachenfeldt, and G. A. Weyhenmeyer. Lakes and reservoirs as regulators of carbon cycling and climate. *Limnol. Oceanogr*, 54(6part2):2298–2314, 2009.

[77] S. Bonaglia, A. Hylén, J. E. Rattray, M. Y. Kononets, N. Ekeroth, B. Thamdrup, V. Brüchert, and P. O. J. Hall. The fate of fixed nitrogen in marine sediments with low organic loading: An in situ study. *Biogeosciences*, 14:285–300, 2017.

[78] D. E. Canfield, A. N. Glazer, and P. G. Falkowski. The evolution and future of Earth's nitrogen cycle. *Science*, 330(6001):192–196, 2010.

[79] S. Pajares and R. Ramos. Processes and microorganisms involved in the marine nitrogen cycle: Knowledge and gaps. *Front. Mar. Sci.*, 6:739, 2019.

[80] M. A. H. J. van Kessel, D. R. Speth, M. Albertsen, P. H. Nielsen, H. J. M. Op den Camp, B. Kartal, M. S. M. Jetten, and S. Lücker. Complete nitrification by a single microorganism. *Nature*, 528(7583):555–559, 2015.

[81] E. Costa, J. Pérez, and J. U. Kreft. Why is metabolic labour divided in nitrification? *Trends. Microbiol.*, 14(5):213–219, 2006.

[82] E. K. Robertson, K. L. Roberts, L. D. W. Burdorf, P. Cook, and B. Thamdrup. Dissimilatory nitrate reduction to ammonium coupled to Fe(II) oxidation in sediments of a periodically hypoxic estuary. *Limnol. Oceanogr.*, 61(1):365–381, 2016.

[83] T. Dalsgaard, B. Thamdrup, and D. E. Canfield. Anaerobic ammonium oxidation (anammox) in the marine environment. *Res. Microbiol.*, 156(4):457–464, 2005.

[84] S. A. Crowe, A. H. Treusch, M. Forth, J. Li, C. Magen, D. E. Canfield, B. Thamdrup, and S. Katsev. Novel anammox bacteria and nitrogen loss from Lake Superior. *Sci. Rep.*, 7(1):1–7, 2017.

[85] C. J. Schubert, E. D. Kaiser, B. Wehrli, B. Thamdrup, P. Lam, and M. M. M. Kuypers. Anaerobic ammonium oxidation in a tropical freshwater system (Lake Tanganyika). *Environ. Microbiol.*, 8(10):1857–1863, 2006.

[86] J. E. Mackin and R. C. Aller. Ammonium adsorption in marine sediments1. *Limnol. Oceanogr.*, 29(2):250–257, 1984.

[87] J. J. Middelburg, K. Soetaert, P. M. J. Herman, and C. H. R. Heip. Denitrification in marine sediments: A model study. *Glob. Biogeochem. Cycles*, 10(4):661–673, 1996.

[88] C. Hensen, M. Zabel, and H. N. Schulz. Benthic cycling of oxygen, nitrogen and phosphorus. *Marine Geochemistry*, pp. 207–240, 2006.

[89] F. A. Roland, F. Darchambeau, A. V. Borges, C. Morana, L. De Brabandere, B. Thamdrup, and S. A Crowe. Denitrification, anaerobic ammonium oxidation, and dissimilatory nitrate reduction to ammonium in an East African Great Lake (Lake Kivu). *Limnology and Oceanography*, 63(2): 687–701, 2018.

[90] A. J. Burgin and S. K. Hamilton. Have we overemphasized the role of denitrification in aquatic ecosystems? A review of nitrate removal pathways. *Front. Ecol. Enviro.*, 5(2):89–96, 2007.

[91] C. C. Michiels, F. Darchambeau, F. A. E. Roland, C. Morana, M. Llirós, T. García-Armisen, B. Thamdrup, A. V. Borges, D. E. Canfield, P. Servais, J. P. Descy, and S. A. Crowe. Iron-dependent nitrogen cycling in a ferruginous lake and the nutrient status of Proterozoic oceans. *Nat. Geosci.*, 10(3):217–221, 2017.

[92] E. K. Robertson, M. Bartoli, V. Brüchert, T. Dalsgaard, P. O. J. Hall, D. Hellemann, S. Hietanen, M. Zilius, and D. J. Conley. Application of the isotope pairing technique in sediments: Use, challenges, and new directions. *Limnolo. Oceanogr. Methods*, 17(2):112–136, 2019.

[93] U. Thomsen, B. Thamdrup, D. A. Stahl, and D. E. Canfield. Pathways of organic carbon oxidation in a deep lacustrine sediment, lake michigan. *Limnol. Oceanogr.*, 49(6):2046–2057, 2004.

[94] J. Li, Y. Zhang, and S. Katsev. Phosphorus recycling in deeply oxygenated sediments in Lake Superior controlled by organic matter mineralization. *Limnol. Oceanogr.*, 63(3):1372–1385, 2018.

[95] B. Thamdrup. Bacterial manganese and iron reduction in aquatic sediments. In *Advances in Microbial Ecology*, pp. 41–84. Springer, 2000.

[96] A. Zegeye, S. Bonneville, L. G. Benning, A. Sturm, D. A. Fowle, C. Jones, D. E. Canfield, C. Ruby, L. C. MacLean, S. Nomosatryo, *et al.*, Green rust formation controls nutrient availability in a ferruginous water column. *Geology*, 40(7):599–602, 2012.

[97] M. E. Jones, J. S. Beckler, and M. Taillefert. The flux of soluble organic-iron (iii) complexes from sediments represents a source of stable iron (iii) to estuarine waters and to the continental shelf. *Limnol. Oceanogr.*, 56(5):1811–1823, 2011.

[98] J. S. Beckler, M. E. Jones, and M. Taillefert. The origin, composition, and reactivity of dissolved iron (iii) complexes in

coastal organic-and iron-rich sediments. *Geochim. Cosmochim. Acta*, 152:72–88, 2015.

[99] J. B. Maynard. 7.11 - manganiferous sediments, rocks, and ores. In Heinrich D. Holland and Karl K. Turekian, editors, *Treatise on Geochemistry*, pp. 289–308. Pergamon, Oxford, 2003.

[100] R. E. Trouwborst, B. G. Clement, B. M. Tebo, B. T. Glazer, and G. W. Luther III. Soluble mn (iii) in suboxic zones. *Science*, 313(5795):1955–1957, 2006.

[101] Andrew S. Madison, Bradley M. Tebo, Alfonso Mucci, Bjorn Sundby, and George W. Luther III. Abundant porewater mn (iii) is a major component of the sedimentary redox system. *Science*, 341(6148):875–878, 2013.

[102] N. Szeinbaum, B. L. Nunn, A. R. Cavazos, S. A. Crowe, F. J. Stewart, T. J. DiChristina, C. T. Reinhard, and J. B. Glass. Novel insights into the taxonomic diversity and molecular mechanisms of bacterial mn (iii) reduction. *Environ. Microbiol. Rep.*, 12(5):583–593, 2020.

[103] C. Hyacinthe, S. Bonneville, and P. Van Cappellen. Reactive iron (iii) in sediments: chemical versus microbial extractions. *Geochim. Cosmochim. Acta*, 70(16):4166–4180, 2006.

[104] S. W. Poulton and D. E. Canfield. Development of a sequential extraction procedure for iron: Implications for iron partitioning in continentally derived particulates. *Chem. Geol.*, 214(3-4):209–221, 2005.

[105] W.K. Lenstra, R. Klomp, F. Molema, T. Behrends, and C. P. Slomp. A sequential extraction procedure for particulate manganese and its application to coastal marine sediments. *Chem. Geol.*, 584:120538, 2021.

[106] A. M. Ure, P. H. Quevauviller, H. Muntau, and B. Griepink. Speciation of heavy metals in soils and sediments. An account of the improvement and harmonization of extraction techniques undertaken under the auspices of the bcr of the commission of the European communities. *Int. J. Environ. Anal. Chem.*, 51(1-4):135–151, 1993.

[107] E. E. Roden and R. G. Wetzel. Kinetics of microbial Fe (iii) oxide reduction in freshwater wetland sediments. *Limnol. Oceanogr.*, 47(1):198–211, 2002.

[108] E. E. Roden and J. W. Edmonds. Phosphate mobilization in iron-rich anaerobic sediments: Microbial Fe (iii) oxide reduction versus iron-sulfide formation. *Archiv für Hydrobiologie*, pp. 347–378, 1997.

[109] Andreas Kappler, Casey Bryce, Muammar Mansor, Ulf Lueder, James M. Byrne, and Elizabeth D. Swanner. An evolving view on

biogeochemical cycling of iron. *Nat. Rev. Microbiol.*, 19(6):360–374, 2021.

[110] O. Larsen and D. Postma. Kinetics of reductive bulk dissolution of lepidocrocite, ferrihydrite, and goethite. *Geochim. Cosmochim. Acta*, 65(9):1367–1379, 2001.

[111] K. H. Nealson and D. Saffarini. Iron and manganese in anaerobic respiration: Environmental significance, physiology, and regulation. *Annu. Rev. Microbiol.*, 48:311–344, 1994.

[112] A. Mostovaya, M. Wind-Hansen, P. Rousteau, L. A. Bristow, and B. Thamdrup. Sulfate-and iron-dependent anaerobic methane oxidation occurring side-by-side in freshwater lake sediment. *Limnol. Oceanogr.*, 67(1):231–246, 2022.

[113] A. Grengs, G. Ledesma, Y. Xiong, S. Katsev, S. W. Poulton, E. D. Swanner, and C. Wittkop. Direct precipitation of siderite in ferruginous environments. *Geochem. Perspect. Lett.*, 30:1–6, 2024.

[114] M. Egger, T. Jilbert, T. Behrends, C. Rivard, and C. P. Slomp. Vivianite is a major sink for phosphorus in methanogenic coastal surface sediments. *Geochim. Cosmochim. Acta*, 169:217–235, 2015.

[115] A. Manceau, M. Kersten, M. A. Marcus, N. Geoffroy, and L. Granina. Ba and Ni speciation in a nodule of binary Mn oxide phase composition from Lake Baikal. *Geochim. Cosmochim. Acta*, 71(8):1967–1981, 2007.

[116] W. E. Dean, W. S. Moore, and K. H. Nealson. Manganese cycles and the origin of manganese nodules, Oneida Lake, New York, USA. *Chem. Geol.*, 34(1-2):53–64, 1981.

[117] J. Li, Y. Zhang, and S. Katsev. Phosphorus recycling in deeply oxygenated sediments in lake superior controlled by organic matter mineralization. *Limnol. Oceanogr.*, 63(3):1372–1385, 2018.

[118] E. Anagnostou and R. M. Sherrell. Magic method for subnanomolar orthophosphate determination in freshwater. *Limnol. Oceanogr. Methods*, 6(1):64–74, 2008.

[119] P. Rimmelin and T. Moutin. Re-examination of the magic method to determine low orthophosphate concentration in seawater. *Anal. Chim. Acta*, 548(1-2):174–182, 2005.

[120] D. L. Correll. Phosphorus: A rate limiting nutrient in surface waters. *Poult. Sci.*, 78(5):674–682, 1999.

[121] A. C. Martiny, J. A. Vrugt, and M. W. Lomas. Concentrations and ratios of particulate organic carbon, nitrogen, and phosphorus in the global ocean. *Sci. Data*, 1(1):1–7, 2014.

[122] R. W. Sterner and D. O. Hessen. Algal nutrient limitation and the nutrition of aquatic herbivores. *Annual Review of Ecology and Systematics*, pp. 1–29, 1994.

[123]   R. W. Sterner and J. J. Elser. Ecological stoichiometry. In *Ecological Stoichiometry*. Princeton University Press, 2017.

[124]   S. Hochstädter. Seasonal changes of c: P ratios of seston, bacteria, phytoplankton and zooplankton in a deep, mesotrophic lake. *Freshw. Biolo.*, 44(3):453–463, 2000.

[125]   J. O. Nriagu. Stability of vivianite and ion-pair formation in the system Fe3 (PO4) 2-H3PO4-H2O. *Geochim. Cosmochim. Acta*, 36(4):459–470, 1972.

[126]   R. Metz, N. Kumar, W. D. C. Schenkeveld, and S. M. Kraemer. Rates and mechanism of vivianite dissolution under anoxic conditions. *Environ. Sci. Technol.*, 57(45):17266–17277, 2023.

[127]   M. Dittrich and R. Koschel. Interactions between calcite precipitation (natural and artificial) and phosphorus cycle in the hardwater lake. *Hydrobiologia*, 469(1):49–57, 2002.

[128]   K. C. Ruttenberg. Development of a sequential extraction method for different forms of phosphorus in marine sediments. *Limnol. Oceanogr.*, 37(7):1460–1482, 1992.

[129]   K. C. Ruttenberg, N. O. Ogawa, F. Tamburini, R. A. Briggs, N. D. Colasacco, and E. Joyce. Improved, high-throughput approach for phosphorus speciation in natural sediments via the sedex sequential extraction method. *Limnol. Oceanogr. Methods*, 7(5):319–333, 2009.

[130]   J. Murphy and J. P. Riley. A modified single solution method for the determination of phosphate in natural waters. *Anal. Chim. Acta*, 27:31–36, 1962.

[131]   D. M. Karl and G. Tien. Magic: A sensitive and precise method for measuring dissolved phosphorus in aquatic environments. *Limnol. Oceanogr.*, 37(1):105–116, 1992.

[132]   M. Dittrich, L. Moreau, J. Gordon, S. Quazi, C. Palermo, R. Fulthorpe, S. Katsev, J. Bollmann, and A. Chesnyuk. Geomicrobiology of iron layers in the sediment of Lake Superior. *Aquat. Geochem.*, 21(2):123–140, 2015.

[133]   S. Katsev and M. Dittrich. Modeling of decadal scale phosphorus retention in lake sediment under varying redox conditions. *Ecol. Model.*, 251:246–259, 2013.

[134]   S. Katsev, G. Chaillou, B. Sundby, and A. Mucci. Effects of progressive oxygen depletion on sediment diagenesis and fluxes: A model for the lower St. Lawrence River Estuary. *Limnol. Oceanogr.*, 52(6):2555–2568, 2007.

[135]   B. Sundby, C. Gobeil, N. Silverberg, and M. Alfonso. The phosphorus cycle in coastal marine sediments. *Limnol. Oceanogr.*, 37(6):1129–1145, 1992.

[136] S. Katsev, I. Tsandev, I. L'Heureux, and D. G. Rancourt. Factors controlling long-term phosphorus efflux from lake sediments: Exploratory reactive-transport modeling. *Chem. Geol.*, 234(1-2): 127–147, 2006.

[137] S. Katsev. Phosphorus effluxes from lake sediments. In *Soil Phosphorus*, pp. 115–131. Taylor and Francis, 2016.

[138] X. Yang, R. Gao, A. Huff, S. Katsev, T. Ozersky, and J. Li. Polyphosphate phosphorus in the great lakes. *Limnol. Oceanogr. Lett.*, 9(5):602–611, 2024.

[139] R. Gächter, J. S. Meyer, and A. Mares. Contribution of bacteria to release and fixation of phosphorus in lake sediments. *Limnol. Oceanogr.*, 33(6part2):1542–1558, 1988.

[140] M. Hupfer, S. Gloess, and H.-P. Grossart. Polyphosphate-accumulating microorganisms in aquatic sediments. *Aquat. Microb. Ecol.*, 47(3):299–311, 2007.

[141] J. Li, V. Ianaiev, A. Huff, J. Zalusky, T. Ozersky, and S. Katsev. Benthic invaders control the phosphorus cycle in the world's largest freshwater ecosystem. *Proc. Natl. Acad. Sci. U.S.A.*, 118(6), 2021.

[142] W. Einsele. Über die beziehungen des eisenkreislaufs zum phosphatkreislauf im eutrophen see. *Arch. Hydrobiol.*, 29:664–686, 1936.

[143] C. H. Mortimer. The exchange of dissolved substances between mud and water in lakes. *J. Ecol.*, 29(2):280–329, 1941.

[144] R. Gächter and B. Wehrli. Ten years of artificial mixing and oxygenation: no effect on the internal phosphorus loading of two eutrophic lakes. *Environ. Sci. Technol.*, 32(23):3659–3665, 1998.

[145] S. Katsev. When large lakes respond fast: A parsimonious model for phosphorus dynamics. *J. Great Lakes Res.*, 43(1):199–204, 2017.

[146] M. Hupfer and J. Lewandowski. Oxygen controls the phosphorus release from lake sediments–A long-lasting paradigm in limnology. *Int. Rev. Hydrobiol.*, 93(4-5):415–432, 2008.

[147] M. Fakhrae, O. Hancisse, D. E. Canfield, S. A. Crowe, and S. Katsev. Proterozoic seawater sulfate scarcity and the evolution of ocean–atmosphere chemistry. *Nat. Geosci.*, 12(5):375–380, 2019.

[148] D. E. Canfield, E. Kristensen, and B. Thamdrup. The sulfur cycle. *Adv. Mar. Biol.*, 48:313–381, 2005.

[149] A. Vigneron, P. Cruaud, A. I. Culley, R.-M. Couture, C. Lovejoy, and W. F. Vincent. Genomic evidence for sulfur intermediates as new biogeochemical hubs in a model aquatic microbial ecosystem. *Microbiome*, 9(1):46, 2021.

[150] C. M. Callbeck, D. E. Canfield, M. M. M. Kuypers, P. Yilmaz, G. Lavik, B. Thamdrup, C. J. Schubert, and L. A. Bristow. Sulfur cycling in oceanic oxygen minimum zones. *Limnol. Oceanogr.*, 66(6):2360–2392, 2021.

[151] B. B. Jørgensen and B. Thamdrup. A thiosulfate shunt in the sulfur cycle of marine sediments. *Science* (New York, N.Y.), 249(4965):152–154, 1990.

[152] B. B. Jørgensen. The sulfur cycle of freshwater sediments: Role of thiosulfate. *Limnol. Oceanogr.*, 35(6):1329–1342, 1990.

[153] F. J. Millero. The oxidation of h2s in black sea waters. *Deep Sea Res. Part A. Oceanogr. Res. Pap.*, 38:S1139–S1150, 1991.

[154] K. Finster. Microbiological disproportionation of inorganic sulfur compounds. *J. Sulfur. Chem.*, 29(3-4):281–292, 2008.

[155] J. P. Amend, H. S. Aronson, J. Macalady, and D. E. LaRowe. Another chemolithotrophic metabolism missing in nature: sulfur comproportionation. *Environ. Microbiol.*, 22(6):1971–1976, 2020.

[156] D. Rickard and G. W. Luther. Chemistry of iron sulfides. *Chem. Rev.*, 107(2):514–562, 2007.

[157] F.-C. A. Kafantaris and G. K. Druschel. Kinetics of the nucleophilic dissolution of hydrophobic and hydrophilic elemental sulfur sols by sulfide. *Geochim. Cosmochim. Acta*, 269:554–565, 2020.

[158] D. Rickard and J. W. Morse. Acid volatile sulfide (AVS). *Mar. Chem.*, 97(3-4):141–197, 2005.

[159] R.-M. Couture, R. Fischer, P. Van Cappellen, and C. Gobeil. Non-steady state diagenesis of organic and inorganic sulfur in lake sediments. *Geochim. Cosmochim. Acta*, 194:15–33, 2016.

[160] G. M. King and M. J. Klug. Comparative aspects of sulfur mineralization in sediments of a Eutrophic Lake basin. *Appl. Environ. Microbiol.*, 43(6):1406–1412, 1982.

[161] M. Fakhraee and S. Katsev. Organic sulfur was integral to the Archean sulfur cycle. *Nat. Commun.*, 10(1):4556, 2019.

[162] M. R. Raven, R. G. Keil, and S. M. Webb. Microbial sulfate reduction and organic sulfur formation in sinking marine particles. *Science*, 371(6525):178–181, 2021.

[163] L. Shawar, I. Halevy, W. Said-Ahmad, S. Feinstein, V. Boyko, A. Kamyshny, and A. Amrani. Dynamics of pyrite formation and organic matter sulfurization in organic-rich carbonate sediments. *Geochim. Cosmochim. Acta*, 241:219–239, 2018.

[164] A. A. Phillips, I. Ulloa, E. Hyde, J. Agnich, L. Sharpnack, K. G. O'Malley, S. M. Webb, K. M. Schreiner, C. S. Sheik, S. Katsev *et al.*, Organic sulfur from source to sink in low-sulfate Lake Superior. *Limno. Oceanogr.*, 68(12):2716–2732, 2023.

[165] M. A. Peña, S. Katsev, T. Oguz, and D. Gilbert. Modeling dissolved oxygen dynamics and hypoxia. *Biogeosciences*, 7(3):933–957, 2010.

[166] I. H. Tarpgaard, H. Røy, and B. B. Jørgensen. Concurrent low-and high-affinity sulfate reduction kinetics in marine sediment. *Geochim. Cosmoch. Acta*, 75(11):2997–3010, 2011.

[167] I. H. Tarpgaard, B. B. Jørgensen, K. U. Kjeldsen, and H. Røy. The marine sulfate reducer desulfobacterium autotrophicum HRM2 can switch between low and high apparent half-saturation constants for dissimilatory sulfate reduction. *FEMS Microbiol. Ecol.*, 93(4):fix012, 2017.

[168] B. B. Jørgensen. A comparison of methods for the quantification of bacterial sulfate reduction in coastal marine sediments. *Geomicrobiol. J.*, 1(1):11–27, 1978.

[169] M. Fakhraee, S. A. Crowe, and S. Katsev. Sedimentary sulfur isotopes and Neoarchean ocean oxygenation. *Sci. Adv.*, 4(1):e1701835, 2018.

[170] E. McKay, S. Katsev, S. Malkin, and T. Ozersky. Widespread occurrence of filamentous Thioploca bacteria in low-sulfate great lakes sediments with implications for sulfur and nitrogen cycling. *J. Great Lakes Res.*, 49(5):1111–1122, 2023.

[171] B. B. Jørgensen and V. A. Gallardo. Thioploca spp.: Filamentous sulfur bacteria with nitrate vacuoles. *FEMS Microbiol. Ecol.*, 28(4):301–313, 1999.

[172] H. Kojima, A. Teske, and M. Fukui. Morphological and phylogenetic characterizations of freshwater Thioploca species from Lake Biwa, Japan, and Lake Constance, Germany. *Appl. Environ. Microbiol.*, 69(1):390–398, 2003.

[173] A. Teske. Cryptic links in the ocean. *Science*, 330(6009):1326–1327, 2010.

[174] H. Fossing, V. A. Gallardo, B. B. Jørgensen, M. Hüttel, L. P. Nielsen, H. Schulz, D. E. Canfield, S. Forster, R. N. Glud, J. K. Gundersen, *et al.*, Concentration and transport of nitrate by the mat-forming sulphur bacterium thioploca. *Nature*, 374(6524):713–715, 1995.

[175] P. G. Appleby. Dating recent sediments by 210 Pb: *Problems and solutions*. Report STUKA–145. Finland. ISBN 951-712-226-8, pp. 7–24, 1998.

[176] P. G. Appleby. Three decades of dating recent sediments by fallout radionuclides: A review. *The Holocene*, 18(1):83–93, 2008.

[177] K. Lång, K. E. Stenström, A. Rosso, M. Bech, S. Zackrisson, D. Graubau, and S. Mattsson. 14C bomb-pulse dating and stable isotope analysis for growth rate and dietary information in breast cancer? *Radiat. Prot. Dosim.*, 169(1-4):158–164, 2016.

[178] A. S. Cohen. *Paleolimnology: The History and Evolution of Lake Systems.* Oxford University Press, 2003.

[179] F. Hjulström. *Studies of the Morphological Activity of Rivers as Illustrated by the River Fyris.* PhD thesis, The Geological institution of the University of Upsala, 1935.

[180] S. Schultze-Lam, T. J. Beveridge, and D. J. Des Marais. Whiting events: Biogenic origin due to the photosynthetic activity of cyanobacterial picoplankton. *Limnol. Oceanogr.*, 42(1):133–141, 1997.

[181] S. Naeher, X. Cui., and R. E. Summons. Biomarkers: Molecular tools to study life, environments, and climate. *Elements*, 18:79–85, 2022.

[182] E. Smeltzer and E. B. Swain. Answering lake management questions with paleolimnology. In *Lake and Reservoir Management—Practical Applications. Proceedings of the 4th Annual Conference and International Symposium. North America Lake Management Society*, pp. 268–274, 1985.

[183] J. Walve and U. Larsson. Carbon, nitrogen and phosphorus stoichiometry of crustacean zooplankton in the baltic sea: implications for nutrient recycling. *J. Plankton Res.*, 21(12):2309–2321, 1999.

[184] H. They Ng, A. M. Amado, and J. B. Cotner. Redfield ratios in inland waters: Higher biological control of c: N: P ratios in tropical semi-arid high water residence time lakes. *Front. Microbiol.*, 8:1505, 2017.

[185] J. J. Elser and R. P. Hassett. A stoichiometric analysis of the zooplankton–phytoplankton interaction in marine and freshwater ecosystems. *Nature*, 370(6486):211–213, 1994.

[186] M. D. O'Beirne, J. P. Werne, R. E. Hecky, T. C. Johnson, S. Katsev, and E. D. Reavie. Anthropogenic climate change has altered primary productivity in Lake Superior. *Nat. Comm.*, 8(1):1–8, 2017.

[187] M. J. Leng and J. D. Marshall. Palaeoclimate interpretation of stable isotope data from lake sediment archives. *Quat. Sci. Rev.*, 23(7-8):811–831, 2004.

[188] H. Graven, C. E. Allison, D. M. Etheridge, S. Hammer, R. F. Keeling, I. Levin, H. A. J. Meijer, M. Rubino, P. P. Tans, C. M. Trudinger, B. H Vaughn, and J. W. C. White. Compiled records of carbon isotopes in atmospheric CO_2 for historical simulations in CMIP6. *Geosci. Model Deve.*, 10(12):4405–4417, 2017.

[189] P. Verburg. The need to correct for the Suess effect in the application of δ13C in sediment of autotrophic Lake Tanganyika, as a productivity proxy in the anthropocene. *J. Paleolimnol.*, 37(4):591–602, 2007.

[190] M. R. Talbot. A review of the palaeohydrological interpretation of carbon and oxygen isotopic ratios in primary lacustrine carbonates. *Chem. Geol.: Isot. Geosci. Sect.*, 80(4):261–279, 1990.

[191] H. C. Craig. The measurement of oxygen isotope paleotemperatures. *Stable Isotopes in Oceanographic Studies and Paleotemperatures: Consiglio Nazionale delle Richerche*, pp. 161–182, 1965.

[192]   D. A. Hodell and C. L. Schelske. Production, sedimentation, and isotopic composition of organic matter in Lake Ontario. *Limnol. Oceanogr.*, 43(2):200–214, 1998.

[193]   P. K. Zigah, E. C. Minor, and J. P. Werne. Radiocarbon and stable-isotope geochemistry of organic and inorganic carbon in Lake Superior. *Glob. Biogeochem. Cycles*, 26(1):GB1023, 2012.

[194]   S. Katsev and I. L'Heureux. Autocatalytic model of oscillatory zoning in experimentally grown (Ba,Sr)SO4 solid solution. *Physi. Rev. E*, 66(6):066206, 2002.

[195]   A. W. Dale, V. J. Bertics, T. Treude, S. Sommer, and K. Wallmann. Modeling benthic–pelagic nutrient exchange processes and porewater distributions in a seasonally hypoxic sediment: Evidence for massive phosphate release by Beggiatoa? *Biogeosciences*, 10(2):629–651, 2013.

[196]   B. P. Boudreau and B. R. Ruddick. On a reactive continuum representation of organic matter diagenesis. *Am. J. Sci.*, 291(5):507–538, 1991.

[197]   H. Shang. A generic hierarchical model of organic matter degradation and preservation in aquatic systems. *Commun. Earth Environ.*, 4(1):16, 2023.

[198]   Q. Jin and C. M. Bethke. The thermodynamics and kinetics of microbial metabolism. *Am. J. Sci.*, 307(4):643–677, 2007.

[199]   E. M. van den Berg, M. Boleij, J. G. Kuenen, R. Kleerebezem, and M. C. M. van Loosdrecht. DNRA and denitrification coexist over a broad range of acetate/N-NO3- ratios, in a chemostat enrichment culture. *Front. Microbiol.*, 7:1842, 2016.

[200]   J. C. M. Scholten, P. M. van Bodegom, J. Vogelaar, A. van Ittersum, K. Hordijk, W. Roelofsen, and A. J. M. Stams. Effect of sulfate and nitrate on acetate conversion by anaerobic microorganisms in a freshwater sediment. *FEMS Microbiol. Ecol.*, 42(3):375–385, 2002.

[201]   P. Berg, S. Rysgaard, and B. Thamdrup. Dynamic modeling of early diagenesis and nutrient cycling. A case study in an artic marine sediment. *Am. J. Sci.*, 303(10):905–955, 2003.

[202]   E. E. Roden and R. G. Wetzel. Competition between fe (iii)-reducing and methanogenic bacteria for acetate in iron-rich freshwater sediments. *Microb. Ecol.*, 45(3):252–258, 2003.

[203]   T.-Y. Ho, M. I. Scranton, G. T. Taylor, R. Varela, R. C. Thunell, and F. Muller-Karger. Acetate cycling in the water column of the cariaco basin: Seasonal and vertical variability and implication for carbon cycling. *Limnol. Cceanogr.*, 47(4):1119–1128, 2002.

[204]   A. J. M. Stams, C. M. Plugge, F. A. M. De Bok, B. H. G. W. Van Houten, P. Lens, H. Dijkman, and J. Weijma. Metabolic

interactions in methanogenic and sulfate-reducing bioreactors. *Water Sci. Technol.*, 52(1-2):13–20, 2005.

[205] D. R. Lovley and M. J. Klug. Model for the distribution of sulfate reduction and methanogenesis in freshwater sediments. *Geochim. Cosmochim. Acta*, 50(1):11–18, 1986.

[206] S. Katsev, D. G. Rancourt, and I. L'Heureux. dSED: A database tool for modeling sediment early diagenesis. *Compu. Geosci.*, 30(9-10):959–967, 2004.

[207] D. Rickard. Kinetics of FeS precipitation: Part 1. Competing reaction mechanisms. *Geochim. Cosmochim. Acta*, 59(21):4367–4379, 1995.

[208] P. Van Cappellen and Y. Wang. Cycling of iron and manganese in surface sediments; a general theory for the coupled transport and reaction of carbon, oxygen, nitrogen, sulfur, iron, and manganese. *Am. J. Sci.*, 296(3):197–243, 1996.

[209] M. D. Krom and R. A. Berner. Adsorption of phosphate in anoxic marine sediments 1. *Limnol. Oceanogr.*, 25(5):797–806, 1980.

[210] L. Yuan-Hui and S. Gregory. Diffusion of ions in sea water and in deep-sea sediments. *Geochim. Cosmochim. Acta*, 38(5):703–714, 1974.

[211] F. J. R. Meysman, B. P. Boudreau, and J. J. Middelburg. When and why does bioturbation lead to diffusive mixing? *J. Mar. Res.*, 68(6):881–920, 2010.

[212] P. Berg, H. Røy, F. Janssen, V. Meyer, B. B. Jørgensen, M. Huettel, and D. de Beer. Oxygen uptake by aquatic sediments measured with a novel non-invasive eddy-correlation technique. *Mar. Ecol. Prog. Ser.*, 261:75–83, 2003.

[213] N. J. Grigg, B. P. Boudreau, I. T. Webster, and P. W. Ford. The nonlocal model of porewater irrigation: Limits to its equivalence with a cylinder diffusion model. *J. Mar. Res.*, 63(2):437–455, 2005.

[214] B. P. Boudreau. A method-of-lines code for carbon and nutrient diagenesis in aquatic sediments. *Comput. Geosci.*, 22(5):479–496, 1996.

[215] A. W. Dale, R. A. Boyle, T. M. Lenton, E. D. Ingall, and K. Wallmann. A model for microbial phosphorus cycling in bioturbated marine sediments: Significance for phosphorus burial in the early paleozoic. *Geochim. Cosmochim. Acta*, 189:251–268, 2016.

[216] P. Reichert and M. Omlin. On the usefulness of overparameterized ecological models. *Ecol. Model.*, 95(2-3):289–299, 1997.

[217] H. Fossing, P. Berg, B. Thamdrup, S. Rysgaard, H. M. Sorensen, and K. Nielsen. A model set-up for an oxygen and nutrient flux model for Aarhus Bay (Denmark). *NERI Technical Report*, 483, 2004.

[218] D. W. Paraska, M. R. Hipsey, and S. U. Salmon. Sediment diagenesis models: Review of approaches, challenges and opportunities. *Environ. Model. Softw.*, 61:297–325, 2014.

[219] A. W. Dale, P. Van Cappellen, D. R. Aguilera, and P. Regnier. Methane efflux from marine sediments in passive and active margins: Estimations from bioenergetic reaction–transport simulations. *Earth Planet. Sci. Lett.*, 265(3-4):329–344, 2008.

[220] D. C. Reed, C. K. Algar, J. A. Huber, and G. J. Dick. Gene-centric approach to integrating environmental genomics and biogeochemical models. *Proc. Natl. Acad. Sci. U.S.A.*, 111(5):1879–1884, 2014.

[221] A. L. Masterson, M. J. Alperin, G. L. Arnold, W. M. Berelson, B. B. Jørgensen, H. Røy, and D. T. Johnston. Understanding the isotopic composition of sedimentary sulfide: A multiple sulfur isotope diagenetic model for Aarhus Bay. *Am. J. Sci.*, 322(1):1–27, 2022.

[222] K. Soetaert, J. J. Middelburg, P. M. J. Herman, and K. Buis. On the coupling of benthic and pelagic biogeochemical models. *Earth-Sci. Rev.*, 51(1-4):173–201, 2000.

[223] I. Markelov, R.-M. Couture, R. Fischer, S. Haande, and P. Van Cappellen. Coupling water column and sediment biogeochemical dynamics: Modeling internal phosphorus loading, climate change responses, and mitigation measures in Lake Vansjø, Norway. *J. Geophys. Res. Biogeosci.*, 124(12):3847–3866, 2019.

[224] S. R. Carpenter. Eutrophication of aquatic ecosystems: Bistability and soil phosphorus. *Proc. Natl. Acad. Sci.*, 102(29):10002–10005, 2005.

[225] R. J. Woosley. Evaluation of the temperature dependence of dissociation constants for the marine carbon system using pH and certified reference materials. *Mar. Chem.*, 229:103914, 2021.

[226] M. A. Menheer. *Development of a Benthic-Flux Chamber for Measurement of Ground-Water Seepage and Water Sampling for Mercury Analysis at the Sediment-Water Interface*. Technical report, US Geological Survey, 2004.

[227] J. Woelfel, A. Eggert, and U. Karsten. Marginal impacts of rising temperature on arctic benthic microalgae production based on in situ measurements and modelled estimates. *Mar. Ecol. Prog. Ser.*, 501:25–40, 2014.

[228] J. Seeberg-Elverfeldt, M. Schlüter, T. Feseker, and M. Kölling. Rhizon sampling of porewaters near the sediment-water interface of aquatic systems. *Limnol. Oceanogr. Methods*, 3(8):361–371, 2005.

[229] N. P. Revsbech. An oxygen microsensor with a guard cathode. *Limnol. Oceanogr.*, 34(2):474–478, 1989.

[230] L. Pischedda, J.-C. Poggiale, P. Cuny, and F. Gilbert. Imaging oxygen distribution in marine sediments. The importance of bioturbation and sediment heterogeneity. *Acta Biotheor.*, 56(1):123–135, 2008.

[231] F. Cesbron, E. Metzger, P. Launeau, B. Deflandre, M.-L. Delgard, A. T. de Chanvalon, E. Geslin, P. Anschutz, and D. Jezequel. Simultaneous 2D imaging of dissolved iron and reactive phosphorus in sediment porewaters by thin-film and hyperspectral methods. *Environ. Sci. Technol.*, 48(5):2816–2826, 2014.

[232] C. Li, S. Ding, L. Yang, Q. Zhu, M. Chen, D. C. W. Tsang, G. Cai, C. Feng, Y. Wang, and C. Zhang. Planar optode: A two-dimensional imaging technique for studying spatial-temporal dynamics of solutes in sediment and soil. *Earth-Sci. Rev.*, 197:102916, 2019.

[233] S. C. Dufour, G. Desrosiers, B. Long, P. Lajeunesse, M. Gagnoud, J. Labrie, P. Archambault, and G. Stora. A new method for three-dimensional visualization and quantification of biogenic structures in aquatic sediments using axial tomodensitometry. *Limnol. Oceanogr. Methods*, 3(8):372–380, 2005.

[234] R. M. Garrels and C. L. Christ (1965). Solutions, Minerals, and Equilibria. Harper & Row (New York), 450 pp.

[235] M. P. Brady, R. Tostevin, and N. J. Tosca. Marine phosphate availability and the chemical origins of life on earth. *Nat. Comm.*, 13(1):1–9, 2022.

[236] R. Sander. Compilation of Henry's law constants (version 4.0) for water as solvent. *Atmos. Chem. Phys*, 15:4399–4981, 2015.

[237] R. A. Berner. Thermodynamic stability of sedimentary iron sulfides. *Am. J. Sci.*, 265(9):773–785, 1967.

[238] A. La Iglesia. Estimating the thermodynamic properties of phosphate minerals at high and low temperature from the sum of constituent units. *Estud. Geol.*, 65(2):109–119, 2009.

[239] F. J. Millero, S. Sotolongo, and M. Izaguirre. The oxidation kinetics of Fe(II) in seawater. *Geochim. Cosmochim. Acta*, 51(4):793–801, 1987.

[240] F. J. Millero, S. Hubinger, M. Fernandez, and S. Garnett. Oxidation of H2S in seawater as a function of temperature, pH, and ionic strength. *Environ. Sci. Technol.*, 21(5):439–443, 1987.

[241] J. W. Morse, F. J. Millero, J. C. Cornwell, and D. Rickard. The chemistry of the hydrogen sulfide and iron sulfide systems in natural waters. *Earth-Sci. Rev.*, 24(1):1–42, 1987.

[242] E. Torres, R. M. Couture, B. Shafei, A. Nardi, C. Ayora, and P. Van Cappellen. Reactive transport modeling of early diagenesis

in a reservoir lake affected by acid mine drainage: Trace metals, lake overturn, benthic fluxes and remediation. *Chem. Geol.*, 419:75–91, 2015.

[243]   J. W. M. Wijsman, P. M. J. Herman, J. J. Middelburg, and K. Soetaert. A model for early diagenetic processes in sediments of the continental shelf of the black sea. *Estuar. Coast. Shelf Sci.*, 54(3):403–421, 2002.

[244]   S. D. Thottathil, P. C. J. Reis, and Y. T. Prairie. Methane oxidation kinetics in northern freshwater lakes. *Biogeochemistry*, 143:105–116, 2019.

[245]   Y. Wang and P. Van Cappellen. A multicomponent reactive transport model of early diagenesis: Application to redox cycling in coastal marine sediments. *Geochim. Cosmochim. Acta*, 60(16):2993–3014, 1996.

[246]   P. Meister, B. Liu, T. G. Ferdelman, B. B. Jørgensen, and A. Khalili. Control of sulphate and methane distributions in marine sediments by organic matter reactivity. *Geochim. Cosmochim. Acta*, 104:183–193, 2013.

[247]   A. W. Dale, S. Flury, H. Fossing, P. Regnier, H. Røy, C. Scholze, and B. B. Jørgensen. Kinetics of organic carbon mineralization and methane formation in marine sediments (Aarhus Bay, Denmark). *Geochim. Cosmochim. Acta*, 252:159–178, 2019.

[248]   J. Rooze, M. Egger, I. Tsandev, and C. P. Slomp. Iron-dependent anaerobic oxidation of methane in coastal surface sediments: Potential controls and impact. *Limnology and Oceanography*, 61(S1):S267–S282, 2016.

[249]   J. Gropp, Q. Jin, and I. Halevy. Controls on the isotopic composition of microbial methane. *Sci. Adv.*, 8(14):5713, 2022.

[250]   B. A. Wing and I. Halevy. Intracellular metabolite levels shape sulfur isotope fractionation during microbial sulfate respiration. *Proc. Natl. Acad. Sci.*, 111(51):18116–18125, 2014.

Index

251

www.ingramcontent.com/pod-product-compliance
Lightning Source LLC
Chambersburg PA
CBHW050550190326
41458CB00007B/1989